ARRIVE

This book is an excellent best-practice guide for senior managers and directors with innovation responsibilities. It describes how organisations of all sizes and sectors can apply design thinking principles coupled with commercial awareness to their innovation agenda. It explains how to keep the customer experience at the centre of innovation efforts and when to apply the range of available practices. It provides a clear, extensive rationale for all advice and techniques offered.

Design thinking has become the number one innovation methodology for many businesses, but there has been a lack of clarity about how best to adopt it. It often requires significant mindset and behavioural changes and managers must have a coherent and integrated understanding in order to guide its adoption effectively. Many design thinking implementations are inadequate or sub-optimal through focusing too much on details of individual methods or being too abstract, with ill-defined objectives. This book uniquely provides integrated clarity and rationale across all levels of design thinking practice and introduces the ARRIVE framework for design thinking in business innovation, which the authors have developed over ten years of practice and research.

ARRIVE = Audit – Research – Reframe – Ideate – Validate – Execute.

The book contains a chapter for each of A-R-R-I-V-E, each of which has explanatory background and step-by-step methods instruction in a clear and standard format.

Using the ARRIVE framework, the book provides high-level understanding, rationale and step-by-step guidance for CEOs, senior innovation leaders, innovation project managers and design practitioners in diverse public and private sectors. It applies equally well to innovation of products, services or systems.

Frank Devitt is an international entrepreneur, engineer and professor. He founded the Department of Design Innovation at Maynooth University, the university's first design activity and Ireland's first university department dedicated to design thinking. He also created and built the award-winning MSc in Design Innovation. Frank regularly teaches, researches, consults, coaches and trains on design thinking, innovation and strategy.

Martin Ryan is a multi-award-winning product designer, entrepreneur and inventor of the Cantilevered Saddle. Awards include: Dyson Design Award (Ireland, 2005) and JEC Composites Innovation Award (Paris, 2016). He is a partner in the consultancy firm, Actionable, which specialises in consumer insight and impactful innovation. He is Director of undergraduate studies in Design Innovation at Maynooth University.

Trevor Vaugh is Principal Investigator in the Maynooth University Innovation Lab (Mi:Lab). Mi:Lab is a Government-funded, design-led lab, formed to explore higher education system challenges. Trevor worked for over 12 years as a medical device designer and holds over 50 patents. Trevor is a partner in the consultancy firm, Actionable, and regularly works with diverse organisations applying design thinking to various challenges.

"The corporate world has long needed – and much anticipated – a thoughtful treatise on the ways in which strategy and design thinking can (and should) work together. With the new book by Devitt, Ryan and Vaugh, I am happy to see that it has finally ARRIVEd!"

—Jeanne Liedtka, Professor of Business Administration, Darden Business School at the University of Virginia

"In a world of hyper change and complexity where being innovative is the difference between success and failure, this book is a must read. For leaders responsible for innovation that need a comprehensive understanding of design innovation, as well as the practical tools to apply, this has it all to ensure your innovations deliver the right value to your users and market again and again."

—Mark Brennan, Vice President Software Development, SAP Labs Ireland

"I really liked your book and can't wait to buy it. It was well worth waiting for it."

—Raomal Perera, Adjunct Professor of Entrepreneurship INSEAD (Paris), Serial Entrepreneur, Consultant

"This is a game changer publication in creating a more practical 'how to' process, building solidly on Design Thinking and packaged as ARRIVE. It offers a new and critical business-strategic alignment as its fourth pillar aligned with human desirability, technical feasibility and business viability. The impact of this makes ARRIVE a powerful tool in mitigating the biggest risks we all have of bringing the wrong idea to market and, equally, failing to change habits and behaviours of our teams."

—Sean McNulty, founder and Chair of the Board at Dolmen Design, Ireland. Ex-Chair of the Technical Committee on Innovation Management at the National Standards Authority of Ireland

"This is an easy-to-read, well-structured, logical and well-illustrated labour of love. It certainly is a very comprehensive tome. As I read through I asked myself would I read this from start to finish or would I use it as a reference book when engaging in a specific piece of work? The answer is that I would use it for both. While the target audience is at a senior level in companies I could see it having a very wide appeal at many levels including people just at the start of their journey. Well done on a lovely publication."

—Stephen Hughes, Enterprise Ireland, Head of Consumer

"ARRIVE offers a memorable and meaningful framework for applying design thinking to innovation projects and embedding a growth mindset in innovative organisations. We've seen it used to guide experienced innovators and newcomers alike as they adopt the new ways of working to put user-experience at the centre of their work. The book describes clearly how ARRIVE helps understanding and guides implementation, as well as including detailed instruction on 35 methods. It's a great resource for all innovation leaders."

—Denis Hayes, Managing Director, Industry Research and Development Group (IRDG)

ARRIVE

A DESIGN INNOVATION FRAMEWORK TO DELIVER BREAKTHROUGH SERVICES, PRODUCTS AND EXPERIENCES

Frank Devitt

Martin Ryan

Trevor Vaugh

Routledge
Taylor & Francis Group

LONDON AND NEW YORK

First published 2021
by Routledge
2 Park Square, Milton Park, Abingdon, Oxon OX14 4RN

and by Routledge
52 Vanderbilt Avenue, New York, NY 10017

Routledge is an imprint of the Taylor & Francis Group, an informa business

British Library Cataloguing-in-Publication Data
A catalogue record for this book is available from the British Library

Library of Congress Cataloging-in-Publication Data
A catalog record has been requested for this book

ISBN: 978-0-367-56662-3 (hbk)
ISBN: 978-0-367-61837-7 (pbk)
ISBN: 978-1-003-10677-7 (ebk)

Typeset in Palatino LT Std
by KnowledgeWorks Global Ltd.

To Heather, for absorbing long days of my distraction and for endless help, advice, support and love.

Frank

I would like to thank my wife Justine for the endless support and freedom she affords me and I hope that Paige (3) and Mason (1) can find some enlightenment in this book when the time comes!

Martin

For Mary, Brody and Lucia, thank you for the endless patience, inspiration and room to immerse, experiment and make a mess.

Trevor

CONTENTS

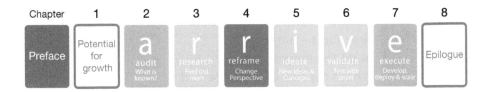

Chapter 1
1

Chapter 6
246

ABOUT THE AUTHORS

Frank Devitt is an entrepreneur, engineer, professor and natural design thinker. Frank has extensive entrepreneurial and strategic management experience across many countries and sectors including industrial electronics, automation, energy and web-based learning. His broader experiences continue to be enriched by regular consulting, coaching and problem solving in diverse industries. Ten years ago, Frank founded the Department of Design Innovation at Maynooth University, which remains Ireland's only university department dedicated to design thinking and business innovation. He also created the multi-award-winning MSc in Design Innovation. Frank's research and teaching focuses on building sustained innovation capability in small and large businesses. Frank rarely refuses a challenge, particularly when it is related to innovation systems improvement and strategic corporate development.

Martin Ryan is Director of Undergraduate Studies in Design Innovation at Maynooth University and lectures in topics of design, entrepreneurship and innovation strategy. Besides his academic responsibilities, Martin is an active designer known for inventing the world's first cantilevered saddle which today is licensed to global saddle brand Stübben. He is also co-founding partner in the consultancy firm Actionable Innovation which specialises in consumer insight, innovation and strategy. For his work in design and entrepreneurship Martin has been acclaimed with many national and international awards from Dyson, JEC World, ISPO and the Institute of Designers, Ireland.

Trevor Vaugh is a lecturer in human-centred innovation at Maynooth University and a founding partner at Actionable Innovation. Trevor merges design, anthropology and business strategy to help organisations develop innovative products, services and strategy. Prior to entering academia, Trevor developed a number of world-first surgical innovations. Among these, his work on single-site surgery for Olympus was awarded a place in the prestigious Cleveland Clinic's top 10 innovations of 2009. Across his career, Trevor has accumulated a portfolio of over 50 patents and co-founded a number of technology ventures. Recently, Trevor appeared as an expert on the critically acclaimed RTÉ television series *The Big Life Fix*.

ACKNOWLEDGEMENTS

This book is the result of over ten years developing our approach to innovation with a design thinking twist. There are a multitude of people who helped along the way, many of them unaware of the assistance they gave. We are especially grateful to Sean McNulty of Dolmen Design and Stephen Hughes of Enterprise Ireland for their helpful comments and suggestions on the penultimate draft of the book. Many others reviewed early stages of the book, in part and in whole, and we thank them all for their assistance in bringing it to a very satisfying conclusion.

All our relations and friends contributed continuously to the creation and maturing of the ideas and structure of the book over the three years it has been in preparation. For example, Frank's daughter Niamh, son Conor and his partner Katy sacrificed peaceful indulgence of a beautiful evening meal in London in September 2019 in order to critique structure, content and some proposed colours. We are all grateful for the support of everyone around us.

In particular, each of us wishes to acknowledge the vibrancy and stimulation of the relationship we have formed together, professional and social. In many ways we are three very different characters. However, we enjoy each other's company, wherein we find a space that enables our diversity to stimulate practical creativity rather than to suppress it. We have been lucky to have spent the last decade or so as professional colleagues.

We are indebted also to IRDG, the Industry Research and Development Group in Ireland, which has formed a very strong and exciting network of practitioners in all things innovation, including specifically design thinking. Through IRDG, and in particular Denis Hayes, Mary Byrne and Bernadette McGahon, we have gained access to many innovation practitioners and businesses in a wide variety of sectors. Together with IRDG we have delivered Design Thinking Masterclasses to almost 100 businesses and over 400 people during the past five years. Through these, we have trialled, validated and incrementally improved many of the techniques we describe in this book.

Another source of development and validation of the techniques we describe here has been our students at Maynooth University's Department of Design Innovation. The three authors have been founding staff members of this relatively new discipline at Maynooth University, and one of the prize jewels has been the award-winning MSc in Design Innovation. We are grateful to the seven years of graduating classes in this MSc programme who have inspired, queried and validated our distinctive ARRIVE methodology. We are also grateful to staff past and present at Maynooth University, for the stimulating critiques and support many have given. We wish especially to acknowledge the support given by our friend of many years, Dr Peter Robbins, also an erstwhile colleague at Maynooth. Through many coffees, occasional dinners and regular joint workshop deliveries inside and outside the university, we always valued the very particular perspective that Peter brought to our thinking.

We wish to thank Maynooth University for allowing us the time to complete this work and to develop the ideas brought together in its pages.

We are grateful to The Explorer's Guide to Biology, Ibiology Inc., MC2240, GHN476D, 600 16th St, San Francisco, CA, USA 94143, for generously granting permission to reprint their image of Francis Crick and Jim Watson with the double helix model.

Finally, we are grateful to our publisher, Routledge, for taking on this project with us and we acknowledge the very easy and productive support of editorial and production staff Rebecca Marsh, Kelly Cracknell and Sophie Peoples, who made it achievable for us.

PREFACE

WHERE FROM AND WHO FOR?

When you ask Frank about design thinking, he says he has been a design thinker all his life. However, he had not heard of the term until about 12 years ago, when a member of the audience at a conference in London asked the speaker "what is design thinking?" Frank thought: "Surely, that's obvious". To Frank, it must simply mean replicating or simulating the way designers think.

So, it still impresses him to realise that, while retaining that elementary understanding, he has spent much of the intervening period reading, research-ing, writing, teaching, coaching and practising design thinking. He has done this in many different contexts including industry, public service, non-profit organisations and higher education. Though he would have done most of these activities in any case, Frank is convinced the explicit understanding of design thinking – as a mindset, set of practices and a process for tackling chal-lenges – has helped him perform and achieve outputs at a higher level than would have been otherwise possible. He has also dealt with a wider range of challenges than he ever could have encountered without design thinking to support him.

Martin and Trevor are product designers who have been practising design thinking for all of their careers in developing magical products, some of which we will encounter later in the book. In more recent years, they have realised how the principles and methods they practised in designing prod-ucts are also applicable to many other areas of life, in particular to business innovation.

In a sense, and from different perspectives, we have come to think of design thinking as a necessary and generic life skill. There is nowhere that it might not be usefully applied, and business innovation is a special case. Further, many of design thinking's elements, such as exploration (curiosity), experimentation (play), and creativity (imagination) are innate characteristics to every human being, but they have been suppressed through traditional education practices and limited framework-based methods of business prac-tice. Our job here is simply to awaken these innate qualities that we all pos-sessed and practised once.

Our mental model for design thinking has flipped between a mindset or a set of practices. Other people consider it just a collection of techniques. Per-haps, according to some commentators, it was just another business fad. Over the years, out of necessity, we have come to see it also as a process.

We know that mentioning the word 'process' sends some creatives and innovators running for cover. But, at its core, a process is simply putting some practices in a certain sequence in order to achieve an output. And, therein lies the concept's value for design thinking. We have seen many attempts at doing design thinking that are either too abstract and lofty or, more often and in contrast, a randomly and incoherently practised collec-tion of techniques. The ARRIVE process that we developed is a process (or framework, or mechanism) for helping innovators understand where they are in the sequence of project activities, why they are there, and what they should do next.

We have found the overview perspective and narrative of the ARRIVE framework brings clarity and better understanding. We do not specify that the implied process must be followed in detail or slavishly. Instead, it provides useful guidance and reference for when alternative sequences or omissions are considered.

We have written this book about the ARRIVE framework for innovation practitioners described in the following pages, people like you.

Joan, Paul, Georgina and Ringo are archetypes of the many individuals we have trained and coached over the last 12 years in particular. In researching our target audience for this book, we created the following persona descriptions. See chapter 4 for more information on using personas.

Joan is VP of R&D in a multi-sited international business in a specialist technology sector. Her main R&D site, which is her personal base, is in a different country from where the business headquarters is located. The CEO, Marketing and Finance functions are based at HQ, so Joan feels isolated. She senses that she is near the bottom of the food chain in her influence on strategic matters of the business. Of course, she and her team's opinions, skills and expertise are highly regarded, but they are 'siloed' and not consulted at the board table as often as Joan would like.

Joan feels her role in innovation, as distinct from R&D, is equally peripheral, being strictly concerned with new product development or new technologies. Usually she is given a brief from Marketing and asked to develop the product to a tight schedule or sooner, with excellent quality or better, and with costs that guarantee target margin or higher. Joan is not expected to go beyond or outside the brief. Any dalliances with customers or key users is barely tolerated, being outside of her core domain. Joan really wants to explore best practices of innovation – for her firm's benefit as well as for her own career and expertise development.

Paul is CEO and majority shareholder in a medium-sized private manufacturing firm, with 120 employees. He knows he needs to shift up a few gears in the innovation stakes in order for the business to survive in an increasingly globalised and competitive market. He wonders how he can manage to do this without interfering in day-to-day business. "My days and weeks are full already!" he explains.

Paul knows innovation always carries a risk, and he wants to contain this at manageable levels. He knows that doing nothing also is a risk. But, he really does not know how to go about being more innovative as everyone seems to be flat out with the ongoing business and there is not a large pool of spare cash available to build up new technology, staff or other capabilities. He is constantly wondering:

"What strategy should I use?"

"What new skills do we need to bring on board?"

"What new product should we prioritise, and for which market?"

"How should I start?"

Georgina has recently been appointed Chief Innovation Officer of a medium-sized multinational consumer goods firm. Despite lots of good wishes and motivating words around the importance of the new role and its pivotal influence on the company's future, Georgina struggles to understand the best way of convincing the board to invest in innovation proposals that are uncertain and risky.

Georgina is well accustomed to making presentations, and she knows that senior executives are better convinced by evidence more than ideas. She feels unconvincing when proposing a bold innovation with inevitably weak 'proof' that it will succeed.

Georgina wishes that she could be able to present a better, more convincing case for the more radical projects that she knows are in the pipeline. It often feels the bolder concepts have flakier, merely qualitative evidence. And, there is no shortage of sceptics. Georgina needs confidence that her methods and approach are best-in-class and solidly grounded.

Ringo is a product designer who has ambitions to climb the managerial ladder in the furniture design and manufacturing firm where he has worked for four years. At the moment he is held in high esteem for producing interesting designs that are aesthetically attractive and functional. But, as time goes by he wishes that he would sometimes be asked to contribute at conception stage, which rarely happens.

Ringo has some allies and supporters among senior management, who sense that he would make a really useful contribution at earlier, more strategic stages in projects. From them, Ringo has learned how to frame his contributions in a more business-relevant way. He needs to learn not just the language of business but also the strategic business concerns of his senior management.

Ringo has recently realised that he might well treat this concern as another design project where, of course, he must empathise with his target user (senior management) and design a way of presenting concepts to them that they feel compelled to support.

Though JPG&R have different professional roles and disciplinary expertise, they are bound by a common desire to know modern, proven and best innovation practices and to learn how they might productively use them in their own work area. Equally, they wish to understand more of the background rationale – the *why* – of the practices recommended here. They want this for their own personal satisfaction and deeper understanding as well as to convince sceptical colleagues and senior management, back at base, of their merits.

It is not a coincidence that the authors of this book have had career roles similar to JPG&R. We have used these diverse experiences in developing our distinctive methods, which we have honed in our engagements with hundreds of firms and many hundreds of individuals from various functional backgrounds over the past 12 years. We are long-term advocates of the principles of Design Thinking and over the years we have augmented established Design Thinking methods with solid, practical business development principles and practices to accommodate questions and circumstances we have encountered from practitioners along the way. What you will find in this book is a robust, sensible, pragmatic, proven and productive approach to realising transformational and sustained innovation.

Of course, we know that not all innovation is transformational, but we know also that transformational innovation is the most difficult and the riskiest. And, it will provide most profits to the business over the next 3–5 years, according to authors Bansi Nagji and Geoff Tuff in their *Harvard Business Review* article of 2012, 'Managing Your Innovation Portfolio'.

In this book, we describe from beginning to end the overall process of designing great, transformational or breakthrough innovation. We explain the rationale for the process and the accompanying suite of tools. Our process is guided by principles based on established best management practice and backed by robust academic research.

However, please do not think that we are fundamentalist in rigidly or inflexibly prescribing how you should apply this framework from beginning to end. On the contrary, we absolutely encourage you to adapt the process to your own circumstances. Use it in full or take parts out of sequence, as you think appropriate. It is rare enough that your context will allow you carte blanche to follow the full, 'perfect' sequence described here. Use it also for non-innovation problem-solving efforts and other business projects, such as designing an innovation strategy itself!

Yet, we structure the book to follow the process of corporate innovation as it naturally unfolds. This forms a compelling and memorable natural narrative. Our main purpose is to illustrate the logic and rationale behind each stage and method as we present it. We show you why the different approaches are taken, which methods are our favourites in different circumstances, and we give detailed step-by-step instructions on using the main methods in the second half of chapters 2 to 7. When you understand more clearly the 'why' and 'how' of the process and its elements, then you are in a better position to choose selectively from it for your particular occasion.

As we write this book, design thinking has become increasingly popular over the last decade in particular. A recent report from PwC consulting company (2018) found that 59 per cent of international organisations identified design thinking as an operating model used to drive innovation within the organisation, second only to open innovation, at 61 per cent. Though relatively new, design thinking has gone mainstream. With that there is a need for it more clearly and effectively to address business imperatives and to contribute to business growth. This imperative is the core mission of this book.

HOW TO USE THIS BOOK

In the introductory chapter 1, Potential for Growth, we give an overview of our approach to design thinking for business innovation, which we call design innovation, and we describe our ARRIVE process framework and rationale. In addition, we introduce our four factors that are the criteria for a successful project, viz. *desirability, viability, feasibility* and *strategic suitability*. The first three are borrowed from existing literature on design thinking. We have added strategic suitability as a necessary additional factor, as we explain later.

After chapter 1, the six chapters 2 to 7 are dedicated to each of the main stages of A-R-R-I-V-E. These chapters have two parts. In the first part, there is an overview narrative explanation of the purpose of the stage and the rationale behind the thinking and practices employed. In the second part, selected methods are explained (mostly) in a standard four-page layout consisting of (1) method overview and step-by-step; (2) template illustration; (3) completed template example; and (4) frequently asked questions.

If you wish to get an overview of the role of design thinking in business innovation, you might read chapter 1 as a stand-alone text. If you wish to understand the philosophy and rationale in more depth, you might also read the first (narrative) parts of each chapter 2 to 7.

Finally, if you wish to study in detail a particular stage of the design thinking process, you might jump straight into the relevant chapter, read the overview narrative and practise some of the methods described there. Design thinking is really a practice of 'doing' and you appreciate its power most when you address its methods to a real, current challenge.

NOTE REGARDING LAYOUT AND WRITING STYLE

We have tried to produce a book that is accessible and useful, in particular to the many practitioners who identify easily with the four target personas introduced above. We have tried to stay away from 'heavy' theoretical discussion or analysis. Yet, we recognise that most people want to believe and are comforted to know that the techniques presented here are solidly grounded in theory and science from a range of disciplines. We have occasionally elaborated on these theoretical underpinnings with a light touch, so that the practitioner can feel more confident to practise the techniques and to defend this practice in the face of criticism from those who are not yet as enlightened.

Regarding gender, the deficiencies of the English language do not allow for easy gender-neutral or all-gender-embracing referencing in everyday language. To overcome this, we have adopted what appears to us the pragmatic, yet fair, approach to all personal pronouns used in the text. In each chapter we use exclusively one gender for pronouns, except where the context demands the other gender. In alternate chapters we use a feminine or masculine gender, respectively, for this generic pronoun. We hope this approach satisfies everybody. We tossed a coin to see which would go first! We start with the feminine in chapter 1.

Potential for growth

Design Innovation – design thinking applied to business innovation – combines deep human understanding, purposeful creativity and sound business rationale in order to create breakthrough innovations that succeed in the marketplace.

ACHIEVING POTENTIAL FOR GROWTH: WHY YOU NEED TO INNOVATE

Hypercompetition is a great word that describes the business conditions for a growing number of industries as the 21st century develops. It refers to increasingly intense and rapidly shifting competition, with regular and unpredictable new entrants creating great difficulty for incumbents to sustain competitive advantage. According to Richard D'Aveni (1998), who coined the term, this environment arises due to four main factors:

- individualisation of customer tastes,
- rapid pace of technological progress,
- globalisation of markets, and
- the rise of giant corporations with deep pockets.

The rules of competition are continuously in flux. Incumbents no longer have power to block new entrants. Barriers to entry are diminishing. There is easier and lower-cost access to sophisticated manufacturing for small firms and individuals; advancements in technology are eliminating cost barriers to developing and testing new offerings, using technologies such as 3D printers, open source CAD packages, software wire-framing and artificial intelligence. Powerful high-street chains no longer control routes to markets; the internet and ecommerce ensure the little guy can appear to the consumer as big and powerful as the incumbent. The small business in Ireland can easily reach markets in Australia, Asia, and the US. Startup businesses are now regularly 'born global'.

    The lesson for business in an increasingly hypercompetitive environment is that traditional competitive strategies cannot be relied on to sustain competitive advantage. Traditional approaches presumed greater stability of consumer preferences, sectoral boundaries, markets and competition. Constant innovation is necessary, not just for attaining new market positions but also to retain the market positions of successful incumbents.

Alongside hypercompetition, the acronym VUCA (*Volatile, Uncertain, Complex and Ambiguous*) is now in widespread use, especially to describe the digital economy. It is a highly disruptive environment characterised by shifting landscapes, which can be punitive to organisations whose focus is merely on incremental innovation.

Kim and Maubourgne's famous book, *Blue Ocean Strategy* (2014), advocates seeking out new spaces in which to do business. These are uncontested markets or blue oceans, where you can offer new value at uncontested prices, instead of trying to compete in the established, highly competitive, 'shark-infested' red oceans. Breakthrough innovation moves the competitive goal posts and changes the rules of competitive advantage to your favour.

There are numerous examples of the rules of competition being changed, through introduction of new technologies or new business models, or both. This has enabled Uber, for example, to become the largest public transport company in the world without owning vehicles. Likewise, Airbnb has become the largest accommodation company without owning property.

Nowadays, more new businesses achieve success through paying very close attention to prospective customers, their behaviours, needs and experiences and through ensuring they design solutions that make the customer's life easier. So, Nest built a business using latest technology to improve the domestic experience of temperature control as well as fire and smoke safety. Nest was founded in 2010 and was acquired by Google in 2014 for a reported sum of $3.2 billion.

While user empathy and understanding are now more commonplace as tools for successful innovation, their use is not an entirely new phenomenon. In the late 1980s, Intuit made life easier for the owners of small businesses trying to manage their business accounts. Intuit's early product Quicken and subsequently Quickbooks spoke to business owners in their own language, rather than as accountants, and enabled them to do more efficiently what they were already doing manually. Intuit has become one of the largest suppliers of business and financial software for small businesses, self-employed individuals and consumers.

Dollar Shave Club was launched in 2012 and provided razors as a service through the post, for a modest subscription price, promoted by a hugely successful viral video campaign. In doing so they helped disrupt the market for men's razors and dislodged dominant incumbents such as Gillette, whose global market share of men's razors and blades fell from a dominant 70 per cent in 2010 to 59 per cent in 2015. In 2016, Unilever acquired Dollar Shave Club for a reported $1 billion.

INNOVATION AND TECHNOLOGY

There is strong statistical evidence that innovation and R&D help make businesses successful. In 2016, the Irish Government Department of Jobs Enterprise and Innovation conducted a study of the economic impact of R&D in firms, covering the years 2003 to 2014. They concluded: "In this study, evidence … indicates a clear correlation between firm engagement in R&D and

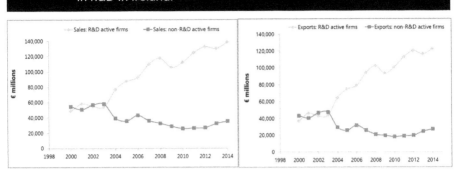

FIGURE 1.1   Economic and Enterprise Impacts from Public Investment in R&D in Ireland.

Source: Irish Government Department of Jobs Enterprise and Innovation, October 2016.

stronger sales, exports, value added and employment performance for both Irish-owned and foreign-owned agency firms". Ireland has shown very strong growth in national economic performance over the decade since 2010, based on its innovation culture.

In Figure 1.1, R&D is assumed to be a proxy for innovation, as is common in many national economic statistics.

However, it is also clear that innovation involves a lot more than just new technology, which was a simplistic default perspective of many people, organisations and governments until recently. The innovation consultants, Strategy&, which is now a division of consulting company PwC, has produced a report every year since 2005, where they study the performance of the top 1000 spenders on R&D among the world's companies. They find consistently that there is no clear correlation between R&D expenditure on the one hand and company financial performance on the other. The conclusion is that R&D alone rarely translates to economic output. This means that, unlike many corporate capabilities, money alone cannot 'buy' successful innovation.

Strategy&'s Global Innovation 1000 report has found the best innovator companies engage in many other complementary activities, besides R&D, that help them consistently outperform those companies focused on R&D spend only. The chart in Figure 1.2 below, adapted from the 2017 report, shows the performance of the top 10 'innovator' companies compared to the top 10 'R&D spenders', across three financial criteria.

WHAT IS INNOVATION?

Innovation is a simple concept. An innovation is something new that is put to use. To be more specific, for an economist, business manager, entrepreneur or designer, innovation is *doing something new that adds economic value through being adopted by a user*. With the same conceptual basis, Intel favours the description 'turning ideas into invoices'.

Note that the term innovation has a variety of meanings for other groups in everyday language. So, an engineer or scientist might refer to innovation as inventing or discovering something new. And an ordinary person reading a newspaper headline might understand innovation to be, simply, anything

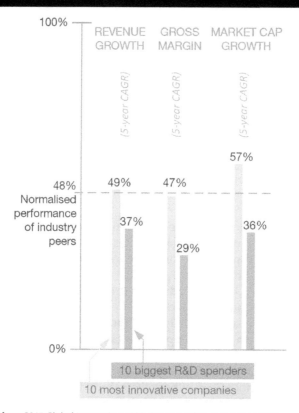

FIGURE 1.2   Comparison of the performance of the Top 10 innovator companies with the Top 10 R&D spenders.

Source: Adapted from 2017 Global Innovation 1000. Strategy& | PwC, October 2017.

new. These are all legitimate in their context, but the latter contexts are not what we discuss here.

In this book, which is a book for business innovation practitioners, our understanding of innovation is always a new value proposition with the following attributes. It is …

	put to use by	→	a **user,**
	provided by	→	a **firm** (or organisation or individual),
	enabled by	→	a **technology** (proprietary or codified know-how),
and	**delivered** to	→	a **market** (that can support a sustainable business model)

There are many different types of innovation. The Oslo Manual (OECD and Eurostat, 2005), which is the handbook for national and international measurements of innovation, describes four key types. They are Product Innovation, Process Innovation, Organisational Innovation and Marketing Innovation. Perhaps more revealingly, the business consulting firm, Doblin, which is now part of the Deloitte company, has famously described ten types of innovation, revised in 2012, as shown in Figure 1.3. These ten types of innovation span across the whole business.

FIGURE 1.3   Doblin's Ten Types of Innovation (2012).

Profit Model	Network	Structure	Process	Product Performance	Product System	Service	Channel	Brand	Customer Engagement

CONFIGURATION   OFFERING   EXPERIENCE

Source: © 2019 Deloitte Development LLC.

DEGREES OF INNOVATION

For every innovation type there are degrees of innovation, from incremental to radical or breakthrough. In the 1980s, the innovation consulting firm Booz-Allen Hamilton (now Strategy&, part of PWC consultants) described six degrees of product innovation. See Figure 1.4, which helps to understand the range that is possible. You will note here, and you will surely have experienced this already, that there are many different terms used for essentially the same or very similar concepts. So, at one end you have incremental, continuous, stepwise or sustaining innovations, and at the other end of the scale you will find discontinuous, radical, disruptive, transformative or breakthrough innovation. Many publications are concerned with parsing fine details of definition to distinguish these different terms. We will not pursue this academic path and in this book we will confine our usage mostly to the broad terms *incremental* and *breakthrough*.

Incremental innovation is something new that represents a seamless improvement from what exists already. The great majority of innovations necessarily must be incremental, as they represent refinement and gradual optimisation of a product, service or system until the scope for improvement has been exhausted, by which time the product and underlying technology becomes a commodity. This may take months, years or decades. The Microsoft Windows upgrade from Windows 8 to Windows 10 was an incremental innovation. Windows 10 introduced many new features and some changed look and feel, but it remains familiar and hence easy to use for anyone familiar with the previous version – and, using it does not offer or require any significant shift in user behaviour.

FIGURE 1.4   Degrees of Product Innovation.

Degrees of Product Innovation (Booz-Allen Hamilton, 1980)

1. Improvements and revisions of existing products

2. New products that provide similar performance at lower cost

3. Existing products that are targeted to new markets

4. Addition of products to an existing product line

5. Creation of new product lines

6. New-to-the-world products

Incremental

Breakthrough

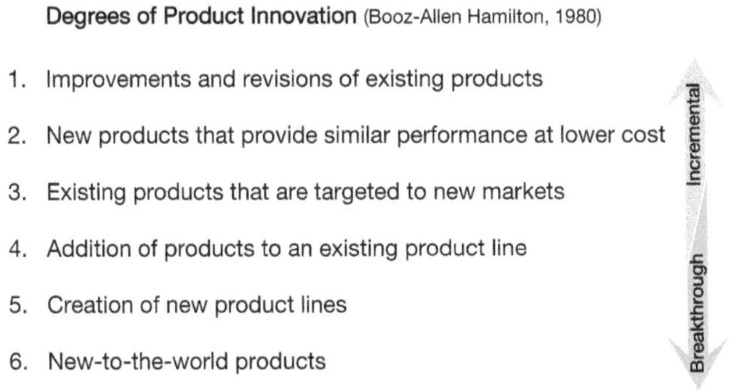

Source: Adapted from Booz-Allen Hamilton.

By contrast, the shift from DOS to the first Windows in 1985 was break-through because it allowed significant and discontinuous shifts in user behaviour and productivity. If you are an Apple aficionado, you will insist (correctly) that the Apple Macintosh was the first commercially successful implementation of a multi-panel window and mouse interface system. It was released in 1984. However, both Apple OS and Windows were heavily influenced by the Xerox Alto computer developed at Xerox PARC research centre in 1973, which was never commercialised! In our worldview, the Xerox Alto was an invention (not put to use commercially); the Apple OS and Windows were innovations!

You might think of incremental innovation occurring when the innovator or user is doing the same things, only better.

Breakthrough innovation, on the other hand, is characterised by the innovator or user, or both, doing significantly different things and achieving substantially better outcomes. To merit the name breakthrough, the difference must be substantial and discontinuous, and is sometimes described as a change in paradigm, i.e. causing a new pattern or model of behaviour and outcomes. The improved outcomes are seen in changes in human experience, new product or system effectiveness, or the opening up of a new market.

It is helpful to understand the categories of innovation by looking at a framework of two axes. On the x-axis is the product or service offering, which describes whether it is existing or new. The y-axis then looks at the target consumer or market. Is this an existing customer segment already served by the firm or is it targeting a whole new constituency? The resulting matrix helps organisations frame both their ambition and the risk that might be involved in a new initiative. Introducing a new product or new technology involves risk, while targeting a new demographic or segment is also risky. Accumulating both risks with a new product and a new market can be so daunting that breakthrough innovation is rarer than it should be for many organisations. In this book, we demonstrate that innovation utilising a design thinking framework provides a path to managing or minimising risk in such projects, although it can never be eliminated.

These concepts have been represented by many authors, in different contexts. In the IDEO terminology represented in the matrix in Figure 1.5(a), the top right, breakthrough quadrant is given the term *revolutionary* (IDEO, 2015). In the Ansoff matrix, represented in Figure 1.5(b), which is regularly used as a strategic planning tool, introducing such an offering is shown to involve a strategy of diversification.

ACHIEVING BREAKTHROUGH INNOVATION

Most of our discussions in this book assume the objective is to achieve successful breakthrough innovation. By successful, we mean enduring commercial success. The innovation is focused on enhancing the experience of users in the target environment, which gives rise to broader market acceptance, thus improving the fortunes of the innovator firm, organisation or entrepreneur.

Initially, the innovator may be inspired by a particular product, service or system of experience delivery that operates in the environment. But, as we shall see, a key aspect of breakthrough innovation is that everything may change and you cannot consider that a physical product must stay as a physical product in the new paradigm, or that any current form of experience-delivery

FIGURE 1.5   Frameworks to represent innovation risk and strategic ambition.
(a) IDEO (b) Ansoff.

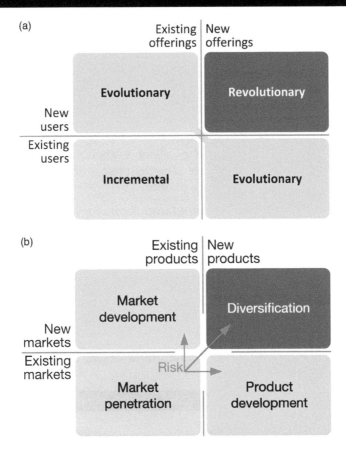

must stay the same. Everything is up for reconsideration and reconfiguration in what we call a 'reframing', a central concept for breakthrough innovation. See chapter 4 for more on reframing.

Note, throughout this book, we use the term product in the broadest sense of physical goods, software package or app, system or service that delivers functional, emotional and experiential benefits to a user within the use environment.

Of course, there are more paths to a successful breakthrough innovation vision than a single technique or methodology will tell. That is not surprising, because innovators come with different styles, shapes and sizes, as individuals or organisations.

Steve Jobs was a confident visionary who looked into his own heart to understand what it was that people needed. Thomas Edison was a logical, incremental, experimental problem solver. Albert Einstein was a creative theoretician who formed a complex and complete theory before bringing his ideas to a practical test.

Google and 3M famously encourage their employees to explore their own individual projects in the '20 per cent time' that is nominally allowed for this, while Apple operates a strong central control on its projects and strategic operating areas.

All of these are successful innovators that have left their marks on the world, but through very different approaches.

Some examples of breakthrough innovators are:

- Southwest Airlines and Ryanair who transformed the experience of air travel for a great part of society through low-cost, no frills, point-to-point flights.
- Airbnb caused large numbers of travellers to abandon traditionally bland, generic hotel accommodation in favour of an earthier, more local, human-encounter experience of staying in another person's home.
- Nintendo's Wii brought gaming into the physical space, creating physical interactions for all the family.
- The Apple iPhone combined excellent aesthetics and user experience design with all the elements of cameras, phones and mini computing to revolutionise all three markets at once.

Often, breakthrough innovation comes from technology breakthroughs, as in applications of GPS positioning information or touch-screen cursor control, although there are often large time lags between scientific discovery, technological development and subsequent commercial application as innovation.

HILL CLIMBING METAPHOR AND BREAKTHROUGH INNOVATION

Don Norman and Roberto Verganti (2014) provide a useful metaphor for innovation, as represented in Figure 1.6. The vertical axis represents product quality, in the sense that the product improves (or fails to improve) the users'

FIGURE 1.6   Adapted representation of hill climbing metaphor for innovation.

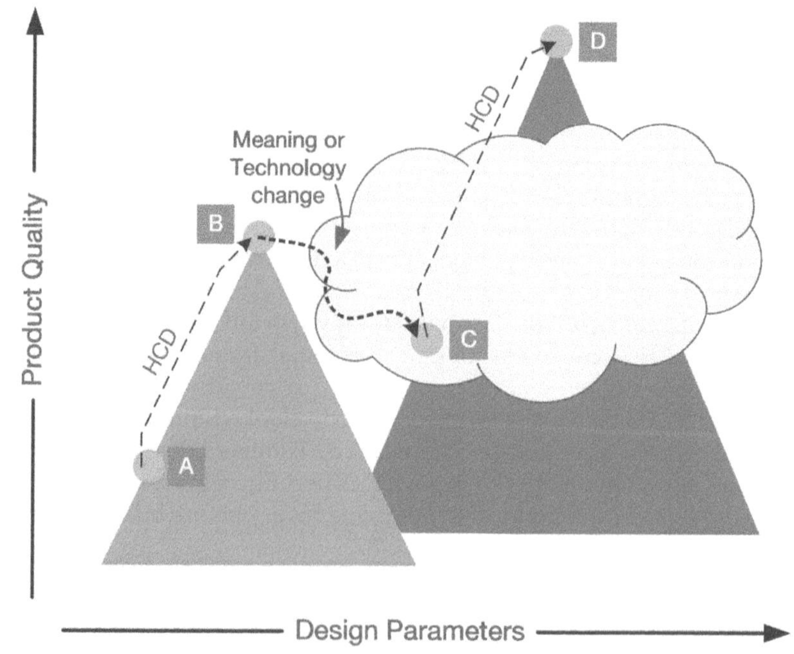

Source: Norman and Verganti, 2014.

experience in the environment. The horizontal axis shows 'design parameters', representing the means by which the product offering is achieved. Incremental innovation occurs from point 'A' when an innovator searches 'locally' for ways to improve the product quality. Such local searching can take you from where you started to a position of incremental improvement, i.e. ascending the hill step by step. Typically, this is done by technology processing improvements or through a study of users that is limited to interactions with existing products or in existing circumstances (the traditional domain of human-centred design, HCD). But, eventually you reach the top of the hill, B, and you can go no further. In commercial terms, the product offering has now been commoditised and there can be no competition except on price alone.

Of particular interest to us, the hill metaphor shows clearly that there may be higher hills remote from the present locality, but the local exploration and test method has no way to access, see or even imagine the higher hill. It takes a breakthrough innovator's keen vision to see through a fog of uncertainty and incomplete data and dare to envisage that other higher peaks may exist at a distance. When conquered, the new higher peaks have the prospect to improve the environment beyond the capability of the present paradigm. In the early stages of this book, we shall focus on cultivating the capacity to form this vision.

Over a cool beer some evening, you might like to ponder the interesting quasi-philosophical point of whether the remote 'higher peaks of breakthrough innovation', described by Norman and Verganti, are waiting to be discovered or to be created. We shall not dwell on this here. For now, we assume that, once the vision is formed, the act of exploring the new vision is one and the same as an act of creation.

Most great breakthrough innovations happen when new technology enables the creation of meaningful new experiences for users, and perhaps creates new users. Norman and Verganti (2014) describe four types of innovation based on changes in technology and meaning, represented in Figure 1.7.

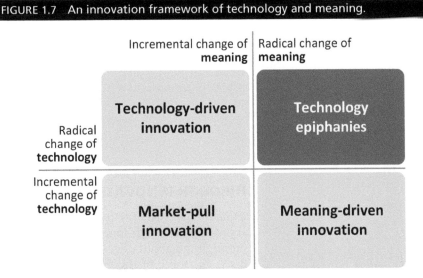

FIGURE 1.7   An innovation framework of technology and meaning.

	Incremental change of meaning	Radical change of meaning
Radical change of technology	Technology-driven innovation	Technology epiphanies
Incremental change of technology	Market-pull innovation	Meaning-driven innovation

Source: Norman and Verganti, 2014.

FIGURE 1.8   A Swatch watch as a fashion item.

The diagram shows that you can have radical innovation with radically new technology or with radically new meaning. The most impactful innovation arises when radical changes in technology and meaning occur simultaneously. In practice, this may act in either direction. In the first instance, meaning change may require enabling technology, either new or existing. Alternatively, a new technology requires a receptive context for it to cause meaning change.

A great example of meaning change can be found in how the Swiss watch company, Swatch, re-invented the Swiss watch industry by moving the meaning of a watch from a time-keeping device to a fashion item. In the late 20th century, low cost micro-electronics and the ubiquity of mobile phones meant that functional, high-accuracy time keeping was easily available and there was little space left for the expensive fine mechanical craftsmanship that had been the basis of the Swiss watch industry for centuries. Reimagining the watch as a fashion item spawned a whole new industry.

Another example of meaning change in breakthrough innovation is seen with the iPhone, which initiated a society-wide radical change in meaning of the 'phone' from its original role as a voice communication tool to its present status as a personal computing device and comprehensive channel of multimedia connection to the external world. Many younger users nowadays rarely use the device as a 'telephonic' communicator!

HOW TO ARRIVE AT BREAKTHROUGH INNOVATION

The key point of interest for practitioners like you, our readers, is how you might come to a vision of such a new breakthrough innovation possibility and, having achieved that, how you might go about developing it to a practicable and profitable offering. This is the topic of this book and we have developed our unique ARRIVE framework to guide you through the various phases that you encounter along the way. Throughout the book, we describe the full ARRIVE framework as you would use it with a full project, from fresh kick-off to full deployment, i.e. from heartbeat to high-street. In fact, ARRIVE is like a meta-process, in that it is compatible with and can be overlaid onto a firm's existing innovation systems and new product development processes.

But of course, in many cases you may be looking simply for extra inspiration in the middle of an existing project. Perhaps, you may wish only to supplement another in-house process in some way, or you may wish to adapt elements of the ARRIVE framework to blend with your own process. We encourage you to do any and all of these in whatever way you feel is appropriate for your circumstances. We do not believe it must be all or nothing in adopting the ARRIVE framework or any other design thinking-based system, because the benefits to be gained are not just about methods and process. The more significant benefits of design thinking are sustained mindset change, increased number of options and validated choices.

> "The more significant benefits of design thinking are sustained mindset change, increased number of options and validated choices."

We refer to ARRIVE as a framework for design innovation (as derivative of design thinking) because our primary goal is to assist practitioners like you to achieve successful business innovation results. We draw substantially from the principles, mindsets and methods of design thinking to help with this and there is now a substantial body of literature and case examples around design thinking that we can refer to. In addition, over the past ten years of our teaching, training and consulting in the space of breakthrough innovation we have developed a clear understanding of how to apply design thinking to real-life projects. We go beyond the sometimes-abstract treatment of design thinking and we provide regular practicality anchors that help blend these to the concerns of real-life organisations and business managers.

Design Innovation is the practice of combining deep human understanding, purposeful creativity and sound business rationale so as to create breakthrough innovations to achieve commercial success. We are careful not to make claims which could be at the same time arrogant, silly and confusing. We do not regard design innovation as a new religion that subverts all previous approaches and methods of innovation. Instead, design innovation and our approach to design thinking is an integrating framework for many of the other methods, theories and approaches to strategic innovation that you have probably encountered. We are happy to show where we connect with and use – and occasionally differ from – many other methodologies such as Blue Ocean Strategy, Jobs-To-Be-Done, Human-Centred-Design, Lean Startup, and other strategic innovation frameworks.

We give a comprehensive overview of design thinking in a later section in this chapter.

The central and pivotal concept of the ARRIVE framework is *reframing*, which is a strategically creative reimagining of what an improved future

FIGURE 1.9 The pivotal point of the ARRIVE process is the Reframe.

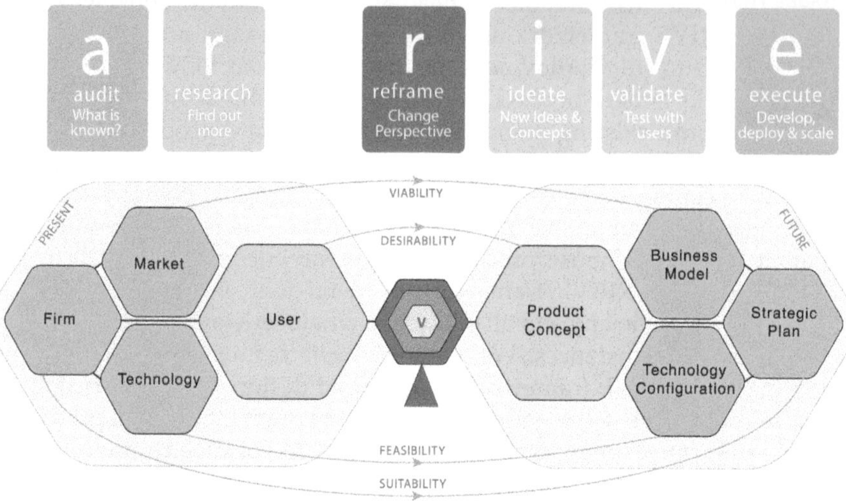

could look like. The Reframed Vision, which is the principal outcome of reframing, is inspired by insights that are carefully extracted from detailed research and recognition of the research output's emergent patterns. You can see this represented visually in Figure 1.9, and it is treated in detail in chapters 2 to 7.

Before reframing, the earlier stages of ARRIVE provide the pathway for the innovator to become embedded in the context of the proposed innovation through ethnography, needs analysis, contextual enquiry, etc. Through this she becomes more deeply informed about the status quo, better able to understand how things are and why they are this way. She develops a deeper empathy with the various stakeholders and users in the system. Through such intimacy and in-depth consideration, she develops the capacity to imagine a future that will deliver significantly improved experiences for the users. For the future concept to be successful in deployment, it must be embodied in a solution that guarantees *desirability* by the targeted users, *viability* as a sustainable business proposition for the marketplace, *feasibility* in respect of available technology or other operational constraints, and *strategic suitability* for the host organisation or individual to undertake the project.

Sometimes, the vision is generated independently by an entrepreneur (often the founder of the firm, the 'hero entrepreneur'), individual domain expert or senior leadership of an incumbent company, and there is little appetite from them for the innovation project team to fundamentally re-evaluate it. The team's job is to act upon it and implement it. See Figure 1.10. We provide more detail on these cases in chapter 4.

We shall elaborate on the ARRIVE process later in this chapter and throughout chapters 2 to 7.

Because the ARRIVE framework for design innovation is closely linked with and derived from the corpus of work on design thinking that has mushroomed over the last ten years, our next section gives a rigorous overview of what design thinking is, and what it is not.

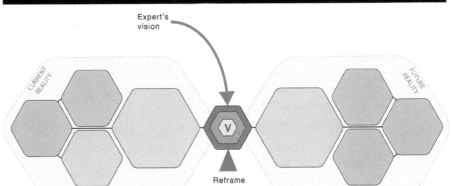

FIGURE 1.10 Expert's vision imposed on the ARRIVE framework.

DESIGN THINKING AS MINDSET

We are often asked: Is design thinking a process or a set of tools? Is it more than both or is it entirely something else? When and why is design thinking useful or necessary?

We aim to give you clear answers and context throughout this book, and we shall remove confusion that may linger after, or even be caused by, the growing volume of literature on the subject.

Here is a good thought to start with. Above all, design thinking is a mindset. That should not be surprising – thinking is in the name! A mindset is a habitual way of thinking that determines how you interpret and respond to situations.

Note, for some practitioners, the presence of the word 'thinking' in the title is off-putting. They see themselves as men or women of action, more Indiana Jones than Socrates. The software giant SAP, a global champion of design thinking for many years, has begun to rebrand the activity in their firm as 'Design Doing' and this may be more appealing to some organisations who find 'thinking' too passive an activity. The name may change, but remember Shakespeare's famous quote: "What's in a name? That which we call a rose by any other name would smell as sweet."

Design thinking is a mindset that restores balance in the way of thinking about problems and opportunities, from excessively rationalistic framework-based thinking that seeks out proof-before-action to enquiry-based thinking that seeks out possibility and proof-through-action. As a mindset, it is a facility that requires, and improves with, practice and experience.

> "Design Thinking mindset shifts from excessively rationalistic framework-based thinking that seeks out proof-before-action to enquiry-based thinking that seeks out possibility and proof-through-action."

The design thinking mindset is particularly useful and attuned to dealing with the uncertainty and complexity found in many real-life circumstances, including the hypercompetitive market environment of 21st century business. This explains its growing popularity over recent decades. It has grown to be a methodology for innovation that is fully embraced by industry, although not always optimally applied.

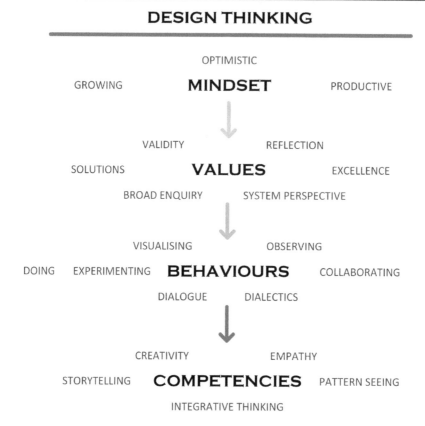

The design thinking mindset is governed by the descriptive set of values you see in Figure 1.11. In turn these values, or operating principles, are operationalised and exemplified by a collection of distinctive behaviours, competencies, methods, skills and tools. We shall elaborate on the mindset of design thinking later. For now, we say that when all these have been systemically diffused throughout the project team and organisation, and are systematically practised to produce something radically new, we can call it a design thinking innovation organisation.

DESIGN THINKING AS PROCESS

Executing a project through the ARRIVE design thinking framework differs in two ways from many processes you may have previously encountered. First, it is less linear and unidirectional. Problem solving progress is characterised by multiple, small-step iterations through reflective cycles of action, learning and re-working.

A second difference is that there is a more fluid approach to planning and implementing the process; there cannot be a strict advance determination of durations for each of the stages because the inherent uncertainty of the context demands adaptability. For example, sometimes the *Research* phase may be especially difficult and complex; at other times, it is prototyping that takes more time, and even then multiple concept iterations may be necessary to confirm a satisfactory outcome. So, the process demands flexibility and attuning to each particular context.

A note of caution: the requirement for flexibility should not be taken as a licence for indolence, avoiding decisions or failing to plan! We have seen this confusion from time to time, not just with design thinking practice but also with other iterative methodologies such as agile software development and Lean Startup. There is a particular skill to be practised in achieving a balance here.

Yet, there is good reason for applying a process description to design thinking. Firstly, it tends to remove some of the mystique that may surround intangible things like mindset and values. Also, it recognises that, in order to achieve useful results using design thinking, intermediate stages must be undertaken and their outcomes captured and used in subsequent stages – all of which defines a process. For example, design thinking demands early-stage research and understanding of the problem space before ideation is attempted. Good quality ideation is founded on deep knowledge of the problem context; it is where the innovator uncovers fresh insights that will unlock creative new ideas and innovative new concepts.

As another example of process sequencing, newly generated solution concepts must be validated with the user and other stakeholders before proceeding to full scale deployment. This does not rule out revisiting an earlier stage on foot of some later stage discovery. After all, continuous refinement through iteration is a distinctive attribute of design thinking. But, it does mandate an overall sequencing of major stages, with early stage outcomes seeding later stages. We are happy to call this a process. See the side box: *Toolbox and process*.

Toolbox and process

While some people do not like to confine design thinking within the bounds of a 'process', we believe there should indeed be a clear sequence of project stages and associated advancement of outputs as a project progresses. This desirable, high-level sequencing is effectively a process, which may be more or less tightly controlled, and more or less fully utilised, depending on the project's particular circumstances.

Think of it this way. A carpenter's toolbox has a variety of tools that the carpenter combines with the trade's methods and skills to produce an artefact. Among the tools may be a measuring tape, saw, sandpaper, screwdriver and paintbrush. To produce a desk or wardrobe, the carpenter acquires the raw material (wood) and uses a measuring tape followed by a saw to produce the sub-components, then sandpaper to smooth the edges, then screwdriver to assemble, followed by paintbrush to finish. These tools are normally used in this sequence; you might call it a process. Still, you can imagine many scenarios where some may be used out of sequence. It is not uncommon to do some re-sanding after a first coat of painting. Or, you might imagine painting some internal parts before full assembly. And so on.

GROWTH MINDSETS AND FIXED MINDSETS

Theoretical support for design thinking's ability to stimulate innovation is found in many areas, not least in the huge literature on the studies of organisational behaviour, organisational learning and developmental psychology, where the idea of mindsets and associated behaviours have been studied.

FIGURE 1.12   The virtuous and vicious cycles of growth and fixed mindsets.

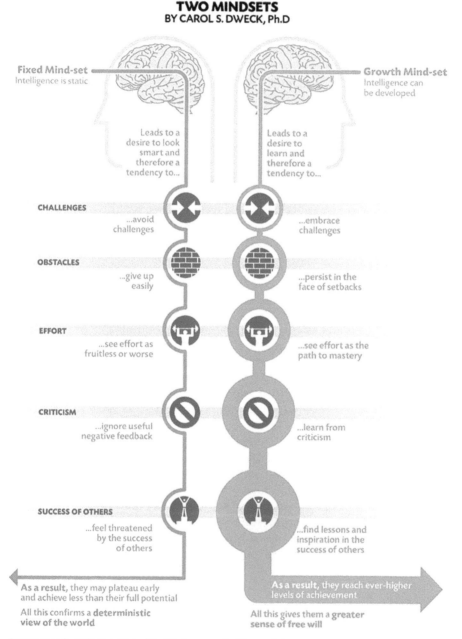

Source: Dweck, 2007. Graphic by Nigel Holmes, used with permission.

Carol Dweck is a professor and researcher in child psychology at Stanford University where she specialises in implicit theories of intelligence and motivation. Over many years of research with children she has identified two distinct characteristic mindsets that children may adopt, which depend on their attitudes to their own abilities, how these attitudes have been nurtured, and how they approach opportunities for learning (Dweck, 2007). Once conditioned to a particular mindset at an early age a person tends to carry this mindset with them through their lives. The good news for many is that mindset may be changed, although with difficulty, through appropriate training or conditioning.

With a fixed mindset, a person regards her own capabilities as determined at birth, and therefore not amenable to development through learning. Circumstances of uncertainty appear as challenges that measure this fixed capability. Consequently, they are perceived as occasions to be feared and avoided.

With a growth mindset, a person regards her own capabilities as amenable to development through new experiences. Circumstances of uncertainty, therefore, become opportunities for learning and are embraced. Learning and development grow in a mutually reinforcing spiral.

Liedtka (2011) has adapted Dweck's model of the two mindsets to describe the stories of two fictional managers, Jeff and George, with growth and fixed mindsets that give rise, respectively, to virtuous and vicious innovation mindsets. This is a very clear exposition of the concept and well worth reading.

A companion theory emanating from Dweck's work is the idea of Goal Orientation and this has been receiving increasing attention in theory and practice. Dweck (1986) postulated that children tend to hold either learning or performance goals. Subsequently, this work was developed and achievement theorists agree that an individual may approach tasks with either of two contrasting achievement goals.

Learning Goal Orientation focuses on task mastery where success is understood in terms of learning new skills and abilities. Individuals define competence self-referentially. They seek learning for its own sake and welcome new challenges.

Alternatively, Performance Achievement goals focus on demonstrating ability and they define competence normatively. Individuals attempt to gain favourable judgements from others, or at least avoid negative judgements. In situations where the latter 'avoidance' behaviour dominates, individuals tend to shy away from challenges.

For our context of designing breakthrough innovation, the Learning Goal Orientation fosters the objective of learning something new and thinking in an unconstrained way. It is more likely to produce better results than simply attempting to keep pace or catch up with what everyone else is doing. It is the difference between playing to win or just to be in the game.

In a business context, Dweck shows that either a growth mindset or a fixed mindset tends to be mutually reinforcing between manager and employees. The manager, holding most power and influence, also holds primary responsibility to drive the direction of this reinforcement and this is achieved by promoting one set of values or another.

Dweck says that a growth mindset supports successful business outcomes in the longer term and also leads to a happier workplace. In a business where a growth mindset flourishes, she describes characteristic behaviours as eagerness to experiment and learn, to collaborate, to consult widely and to openly dissent and argue around a proposal.

In Dweck's model, contrasting with growth mindset, a fixed mindset in business assumes there is a known correct way of doing things, usually the way endorsed by the boss. Dissent is culturally discouraged. Individual genius is overly admired and venerated; hence, wide consultation is eschewed. Success in these companies, if it may be achieved at all, is usually short-lived.

PRODUCTIVE MINDSETS AND DEFENSIVE MINDSETS

Dweck is not alone. Chris Argyris was among the founders of the discipline of Organisation Behaviour and a contributor to many of its theoretical pillars. He also was an expert in organisational learning and in how people produce action that is effective, with consequences that persist (Argyris, 1983, 2004; Argyris and Schön, 1978).

Argyris uses the term Governing Values for the underlying values people use to design their actions. In his theory there are two categories of Governing Values, which give rise, respectively, to a *Defensive Mindset* and a *Productive Mindset*. Though arising from a distant stream of research, these concepts have a lot in common with Dweck's fixed and growth mindsets.

Argyris' Defensive Mindset's governing values essentially are default values for most people and they are:

1. To be in control of one's environment
2. To seek to win and not lose
3. To suppress negative feelings
4. To be as rational as possible

The primary action strategy that arises from these values is to advocate your position and then bully others into agreeing with your way of thinking. Argyris (1983) says: "The fundamental consequence of all this is that people withdraw. They play it safe and try not to make waves. In this environment, people hesitate to be honest, and the undiscussable gets covered up."

Turning to the *Productive Mindset*, its governing values are:

1. To obtain valid information
2. To create conditions for free and informed choices
3. To accept personal responsibility for one's actions

Argyris says that action strategies arising from these values combine *inquiry* and *public testing*. They are *collaborative*, *reflective* and *self-correcting*. He says that any environment that rewards participation, joint problem solving and openness can be expected to move towards a Productive Mindset.

DESIGN THINKING BEHAVIOURS AND VALUES

So, what do the theories of Dweck and Argyris mean for design thinking?

The positive mindset messages of Dweck, Argyris and others emphasise primacy of truth – in the form of validity – and broad enquiry with open discussion in search of that truth. These two values are also at the heart of design thinking. You will find them referenced variously as *validity, deep understanding, human-centredness, co-creation* and *experimentation*. This strong alignment in the fundamental values is why we say that *design thinking is a mindset for growth and productive problem solving.* Design thinking introduces positive behaviours and values, and re-orients the mindset of individuals and organisations.

Validity vs reliability

Reliability is an attribute concerned with consistency and predictability.
   Validity is concerned with appropriateness and alignment of actual with intended outcomes.

FIGURE 1.13   Design thinking is a mindset for growth and productive problem solving.

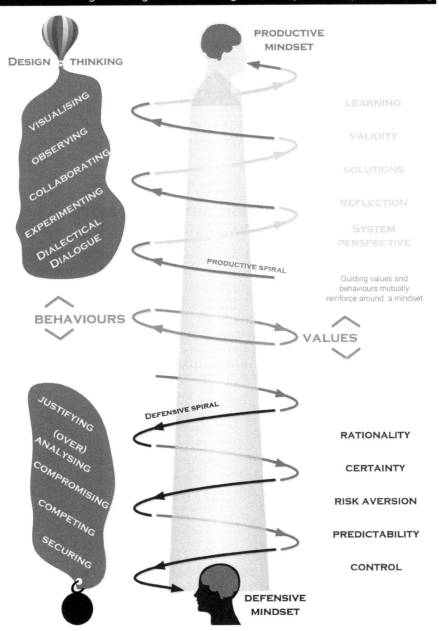

Both Dweck and Argyris refer to an internal reinforcing loop connecting mindset, values and actions. When that loop is spiralling in the negative (defensive or fixed) direction, design thinking can be a powerful intervention into the cycle, which serves to change its orientation towards productivity and growth. Its power comes particularly from its emphasis on *visualisation, observation, collaboration* and *experimentation*, all of which are positive externalised actions that over time directly influence values and indirectly impact the mindset. Individuals, through their repeated actions, become habituated to the positive growth orientation and the mindset is gradually reprogrammed. In turn, changes to individual mindsets gradually bring about cultural change in the whole organisation (Lehman et al., 2004).

The most difficult step in a defensive spiral intervention is the first one. Scepticism and fear are powerful inhibitors. The individual's scepticism comes from poor understanding and lack of experience with the new approach and the new values. Fear of failure, the scorn of peers (Performance Orientation) and the reproof of bosses are valid concerns where new behaviour is significantly different to that which is embodied in the existing culture. So, the first step should be designed to achieve 'quick wins' through which the participants will gain confidence while others, though perhaps sceptical, at least will have curiosity aroused and perhaps be more tolerant for the second step.

DESIGN INNOVATION: DESIGN THINKING ADAPTED TO PRACTICAL BUSINESS INNOVATION

Notwithstanding its tremendous power as a medium for mindset change, design thinking is most directly useful for facilitating breakthrough innovation and creatively solving complex problems. Both of these contexts are characterised by a current reality that is not ideal; the influencing conditions are only vaguely understood; even the nature of an improved future is unknown or only vaguely specified. Furthermore, the pathway to achieve a better future is even more uncertain. These are so-called *wicked* problems, as first described by Rittel and Webber (1973) and elaborated by Buchanan (1992) in a design context.

Wicked problem

A puzzle so complex, persistent, pervasive or slippery that it seems insoluble. There are difficulties of problem definition as well as solution identification. The mindset must be to bring about *an improvement* in the current situation, rather than attempting to find a non-existent unique, optimal solution (Rittel and Webber, 1973).

Over ten years training, teaching, researching and mentoring design thinking to experienced practitioners and postgraduate students, we have encountered a great variety of problems and innovation challenges, with varying degree of difficulty or wickedness. Through hundreds of workshops, training and coaching sessions, we have developed and refined our design thinking methodologies into a framework that guarantees creatively distinctive and practicable outputs for any business innovation project.

Our Design Innovation approach is a practical adaptation of design thinking and encompasses two paradigms, four dimensions and a process with six stages, under the rubric of ARRIVE, as we have shown earlier in Figure 1.9, and repeated in Figure 1.14 below.

> "Everyone designs who devises courses of action aimed at changing existing situations into preferred ones." (Herbert Simon, Nobel laureate in Economics, 1978, and founder of the field of Artificial Intelligence)

The two paradigms of the ARRIVE framework are the current and future realities that were envisioned by Nobel laureate Herbert Simon. The current reality is often truly 'wicked'. It is non-ideal, maybe extremely so, and it is often not easy to access the influencing conditions and the key stakeholders. It is not clear what the route to a satisfactory solution might be and even less so what the nature of a satisfactory solution may be. In order to understand the current reality paradigm, we must examine its four key dimensions. And, in order to create the improved future, we must create an innovation that has the corresponding four required improvement attributes. These dimensions and attributes are represented in the innovation definition that we introduced at

FIGURE 1.14 The ARRIVE framework describes the current reality in four dimensions and maps each dimension to an improved future via innovation attributes.

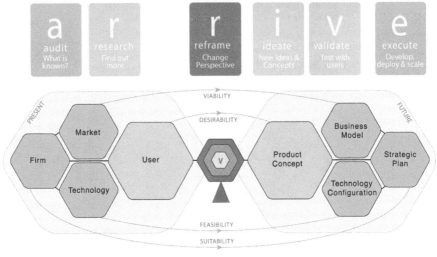

© 2017 Devitt Vaugh Ryan

the beginning of this chapter. That is, the innovation we seek is a new value proposition that is:

	put to use by	→	a **user**	because it is	→	**desirable**
	provided by	→	a **firm**	because it is	→	**strategically suitable**
	enabled by	→	a **technology**	that makes it	→	**feasible**
and	delivered to	→	a **market**	with a	→	**viable** business model
			environmental dimension			*innovation attribute*

Note the first two lines above are about *why* the innovation will be successful, while the second two lines are about *how* the innovation will be successful. Note also that *user* and *market* are concerned with how the organisation interacts with the *external* world. In contrast, *firm* (innovator) and *technology* are focused on *internal* capabilities, strategic and technological. The Four Dimensions of Innovation are mapped in Figure 1.15, and we elaborate on them in the next section.

Three of the four attributes of a successful innovation that we have described above are the attributes of good design that Tim Brown (2009) identified in his book *Change by Design*, which are now widely referenced in the design thinking literature. Brown called the attributes *desirability, viability, feasibility* (and we have followed this) and these relate to our dimensions of *user, market, technology* respectively.

FIGURE 1.15   Four dimensions of innovation.

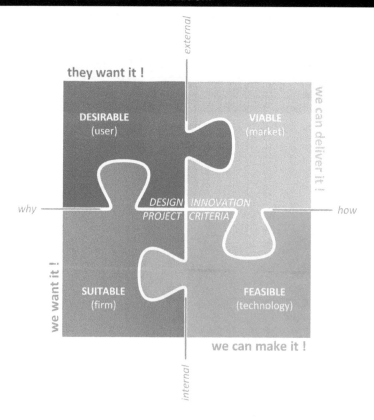

Through our engagements with numerous clients, we found it necessary to add the fourth dimension, the *(innovator) firm*, with an attribute of *strategic suitability*.

THE FOUR DIMENSIONS OF INNOVATION

User (target for *desirability* attribute):

The most significant innovation trend over the last half-century has been to recognise that the user (or a proxy) pays the bill. User behaviour and user experience are key factors in the success or failure of any innovation; hence, innovation must be user-centred as a priority. That same user is now more educated, discerning and demanding than ever before. As well as functional needs to be satisfied, users (and other stakeholders) have emotions and aspirations that drive their behaviours. Problems – or innovation opportunities – arise when users' negative emotions are triggered in any circumstance or when aspirations are not met.

Note that mechanisms and machines do not (yet) have emotions. Strictly, only people have problems. Even in the 21st century, it is hard to define a problem or a potential improvement that does not involve people at some level. Frank sometimes makes a bet with his audience that they cannot find such a problem, and he has not yet parted with his money!

Many of the more complex wicked problems have human behaviour, emotions, experiences and aspirations at the very centre. Humans' variability, apparent irrationality (not as often as is thought) and unpredictability make them all the more challenging to work with in problem solving and exploring innovation. Yet these characteristics make the search for their potential solutions more intriguing.

All design thinking's solutions are derived from deep user research and they focus on enhancing the experiences of the user.

Note, in the terminology of physical products and services we call them users, but 'users' in this book should be understood to mean all stakeholders who are affected by the problem, will be affected by the solution and who will influence its success in the market. It is the users who provide the core understanding of the problem space and who will determine the degree of acceptance of the solution.

Therefore, understanding the users, their experiences, and the context of use is supremely important for innovation. We spend time and try hard to understand the deeper fulfilment needs and aspirations of our prospective users and we keep digging to unearth new insights that may be inspiring though perhaps subtle. The insights can be valuable levers for delighting the users with unexpected new levels of fulfilment, which is the basis of successful market differentiation.

Insights

Insights from user research are nuggets of special, deep understanding about an individual's or group's needs, aspirations, behaviours, attitudes or emotions, as well as societal influences and trends.

Insights are non-obvious, maybe intuition based, discoveries that lead to a new mental model.

Market (target for *viability* attribute):

The people-oriented *desirability* dimension focuses on the target user or stake-holder and seeks to compel the user to acquire the proposed solution by the promise of its improved experience offering. Market considerations go further and in the first instance ask if there are enough such prospective users and a satisfactorily efficient and effective business model to support a sustainable business. It asks and answers the following questions to establish a level of viability for deployment to the market:

- What is the accessible market size?
- What are the individual/sub segments within it?
- How can prospective users be accessed?
- How can they be encouraged to buy? Or, in a non-commercial con-text, how can they be encouraged to participate in or use the proposed solution?
- Can the product or service be deployed at satisfactory levels of capital investment and recurrent costs?
- Is there an effective and sustainably profitable business model?

Technology (target for *feasibility* attribute):

The *feasibility* dimension establishes a technological configuration that will implement a proposed solution and that is compatible with the market, pro-posed business model and general context of production and use. The big question is, what know-how and technical configuration is necessary to sup-port deployment of the proposed solution?

We include in the term *technology* all skills and specialist know-how that must be brought to bear for the deployment of the proposed solution or inter-vention, i.e. not just engineering or scientific skills and know-how.

There are multiple levels of technological feasibility to be assessed, as follows:

- Is there a technology available (anywhere in the world) that would make the proposed solution feasible?
- Is this technology available to the firm or project group implementing the project?
- Is the time frame and expense to operationalise the technology acceptable?
- Is the development risk acceptable? This depends on the level of explor-atory research (high risk) compared to incremental technology develop-ment, repurposing or application (lower risk).
- Does the firm or project group have the capacity to absorb and manage the technology?
- Is the technology suitable for the environment-of-use and for the user?

When the answer to each question above is positive then we can say that the proposed solution is technologically feasible.

Firm (innovator organisation or individual, target for *strategic suitability* attribute):

Nearly every project has a context of origin, an environment where it is rooted and is expected to develop and grow. It may be a particular business, a business division or a startup enterprise defined by the mind, resources and milieu of an entrepreneur. Very few projects are so unfettered that they are free to land in the ideally prepared set-up for development and deployment.

While desirability, feasibility and viability describe a project with intrinsic virtues and prospects for success, in real life the proposed project must be nurtured, massaged, promoted and supported in amenable internal and external environments where it can find its feet and grow. It must be capable of being put into practice in this environment. That is, it must be *practicable* and *suitable* for the environment! It is like a healthy oak sapling that may be viable, feasible and desirable in fertile soil, while its development is practicable only when situated in a suitable environment with the right climate, receiving the right nurturing and protection to allow it to grow to its full potential.

In addressing the environment where a new project is to be nurtured, a wide range of business study parameters could be considered, e.g. organisational structure, culture, leadership and management, strategic development, dynamic capabilities and much more. In fact, everything about the business will contribute to the project's success or failure. However, even for most organisations that are well run for everyday business, there are particular challenges that must be addressed when proposing a new breakthrough innovation project. The main innovation challenges to be addressed are:

- Is there a *mindset* for growth, with the necessary aptitude for exploration and learning that will allow the project to seed and grow?
- Is there a will to proceed? Is the leadership and general membership psychologically ready to take on a project of this nature? It may be too ambitious. Morale may be low after a long period of disappointing performance and huge survival effort. Or, there may be a sense of complacency after a prolonged period of easy and profitable growth. Very often a radical innovation project needs a champion who has power and influence to bring about a sufficient level of enthusiasm and support in the organisation. Certainly, the business leadership needs to be fully behind the project, supporting and managing the necessary and sometimes profound changes that must be brought about.
- Are there sufficient levels of necessary resources available, and can they be ring-fenced to ensure timely project completion?
- Is there alignment with the high level strategic intent of the business?
- Does the organisation's capability for timescale-to-deployment fit the project's requirements?

THE ARRIVE PROCESS FOR DESIGN INNOVATION

Design thinking provides the mindset, behaviours and tools that are useful for solving difficult challenges. However, innovators who use design thinking are not magicians; a design thinker does not achieve excellent outcomes by power of aura, by just being physically present or mentally attentive. Nor is it likely that a design thinker will act by a randomly sequenced series of actions or random selection from the repertoire of tools.

Instead, each design thinker will have her own process, a sequencing of actions, artefact creation, interpretations and staging points that repeatedly construct the routes to her designed outcomes in different problem areas. Designers and problem solvers may differ, as will their process details, but a lot has been learned in the past 20 years about the common elements of the best-in-class of such processes. The ARRIVE framework builds on this knowledge with a six-phase generic process, as depicted in Figure 1.14. We elaborate on each of the six stages in the following sections.

A design innovation project starts with an acknowledgement that innovation opportunities always occur within a context that has multiple actors, structural complexities and legacy momentum. The innovator's starting point must be to find out as much as possible about this environment, much of which probably already has been studied. Probably, the innovator has some knowledge, even expertise in the subject area, but it is rare that a fresh broad review of up to date information does not add value. A lot can be achieved by diligent desk research alone, accessing public information about the target markets, industry sector and key players. Specialist market reports, socioeconomic trend and foresight reports may be complemented by speaking with experts and accessing any other generally available sources.

In considering innovation opportunities, we must investigate the status of our four innovation dimensions: user, market, technology and firm. Traditional business strategy has a variety of tools and techniques to achieve this for the latter three dimensions, and these are what we focus on in the Audit phase. We are setting the scene, so to speak. Some examples are: SWOT analysis, Force Field analysis, Porter's Five Forces of Competition. All of these are introduced in chapter 2, Audit. Understanding the *user* is the focus of the Research phase of chapter 3.

More recently, Robert Verganti (2016), in his book *Overcrowded*, has described a method of 'criticism' for distilling the expertise that usually exists within an established firm. This process of 'criticism' may become the starting point for innovation efforts, skipping out on additional auditing and research described here, and producing an innovation vision as direct output from the criticism process. This is also introduced in chapter 2.

When you have mined all known existing information about the innovator/firm, the clear competition and the market, certainly you are better informed but it is most likely you are not yet ready to create a breakthrough innovation. You might be able to come up with random or obvious ideas, and some indeed may have potential. But such innovation ideas are based on data that are mostly also available to others. It is really hard to be distinctive and differentiated in the market when working from the same knowledge base as competitors about the most important dimension and central focus, the user.

The prospective users and other stakeholders must be identified. You must understand very well their behaviours, and causes of behaviours, as well as their aspirations for deeper fulfilment and better outcomes. You must search deeply for hitherto hidden nuggets of understanding, new insights, that you can leverage to create a distinctive innovation offering.

Reframing is the *secret sauce* of design innovation. The innovator comes to look at the innovation environment with a new perspective, or worldview, that is formed and informed by the deeper understanding and insights learned from research. While the deeper understanding of the context and its actors helps shape the new frame, so too the new frame defines a new scope for the upcoming conceptualisation and development of the innovation.

Innovations cannot be all things to all people. The frame chosen after research should be the one most likely to provide a compelling solution for the main users, which addresses the key issues experienced by them and offers greatest impact potential. Reframing becomes possible only after a period of deep research, understanding and recognition of patterns of data that lead to new insights.

A good example of reframing is the introduction by Philips of its *ambilight* TV in 2004. The new concept came from the insight that a TV is not just a viewing screen in most people's homes. Instead, the ambilight applies a new frame where the TV is offered as a central piece of furniture in the living room, which is the room where the main TV is located in most homes. The innovation was to create a TV that could act as a lighting source to set the mood of the whole room in sympathy with or even independent of the content of its programme delivery function.

r

reframe

Change Perspective

FIGURE 1.16   Philips ambilight TV reframes the meaning of a TV in the living room.

Having reframed the perspective and scope of the innovation problem being addressed, aided and supported by fresh insights, the innovator has an inspiring new vision of an attainable future. With a clear focus and renewed purpose from the reframed vision, the team changes mode and starts to explore the range of possible solutions. Ideation always begins with an exploratory, divergent phase capturing rekindled, new and possibly wild ideas, that might be candidates for further consideration or intermediate stepping stones to a great new concept. To help with ideation, there is no shortage of idea generation methods and in chapter 5 we recommend our favourites. At this early stage, remember that plentiful variety is a goal and team collaboration is essential to ensure diversity of ideas. Quality of ideas will be honed by filtering, selecting and iterating, at the subsequent concept stage.

Here, we follow the dictum of Linus Pauling, the only winner of two unshared Nobel Prizes, for Chemistry (1954) and Peace (1962), who said: "The best way to have good ideas is to have lots of ideas".

Having generated lots of idea 'nuggets', now the innovator must filter, classify, evaluate these and converge on a small number of fleshed out, purposeful, tangible concepts. The concepts are represented visually at a high level by sketched images, annotation notes and perhaps low-fidelity card or video prototypes. Low-fi prototypes and concept boards are continuously and iteratively checked and discussed with users to be sure that the resulting concept is compellingly attractive and shows real promise to help the user achieve progress or resolve a pain in her life.

The concept describes the essence of the proposal and should not be obscured by the finer implementation details, which are for later. Tom Kelley of IDEO calls it "mastering the art of squinting". It is important to be very clear about the core value proposition to the user and the critical attributes that must be in place for it to succeed.

The outcome of the ideation stage is one or two strategically outlined concepts together with identified assumptions behind their case for success. It is these assumptions (together with others) that require validation, or correction, before proceeding to further development.

Until now, design innovation's human-centred approach has given priority of focus to the user dimension of the innovation. You have verified the desirability attributes of your selected final concepts through low-fi prototyping and constant checking with users.

From here on, as well as talking about a *concept* (a term concerned mostly with desirability), we progress to thinking more about a full *business proposal*, encompassing all four dimensions of user, technology, market and firm. The business proposal elaborates on all key factors required for success of the project. Many of these success factors are in the form of assumptions that add to the desirability assumptions of the basic concept.

The objective of the Validate phase is to test and confirm – or alter, replace or abandon – these critical assumptions. As they become more solidly founded, everyone's confidence grows that the proposal really will reach expectations after it has been fully developed and deployed. In this manner, de-risking is the focus of the Validate phase.

Validation is achieved by offering prototypes to the users, and other relevant stakeholders, while observing and evaluating the users' responses. A prototype is anything that allows the user to respond to some tangible experience of the new system, and allows the innovator to learn from this response and develop the concept by iterative, incremental refinement. Sometimes a full system prototype may be appropriate, albeit usually at reduced fidelity. At other times, a specific assumption may be isolated and represented in a partial system prototype or in a separate experimental arrangement constructed solely for that purpose.

> **Prototype**
>
> Any artefact that may be used in an experiment to allow the user to respond to a tangible experience of the proposed concept, or concept element, and allows the innovator to learn from this response.

At early concept stages, prototypes are often low-fidelity and rapidly constructed in the form of sketched storyboard, cardboard model, 'Wizard of Oz' video, and many other methods. The possibilities are limited only by imagination, and there is great scope for exercising creativity to devise a prototype that will provide maximum learning at minimum cost. As the innovation becomes more developed, later-stage prototypes will evaluate user responses to specific features in fine detail or may test global system performance and user response to a full system experience.

To give an example of a low fidelity prototype, David and Tom Kelley (2013) in their stimulating book, *Creative Confidence*, describe the early stage concept prototype that was prepared in one hour by two app designers (Adam Skaates and Coe Leta Stafford) working on a project with Sesame Workshop. The iPhone app was called Elmo's Monster Maker and they were considering a feature where a child user would be able to control the movements of the 'monster' by touching the screen. But, how could they adequately explain this intended feature to their clients on an upcoming phone call?

Within one hour, they cleverly simulated the feature using a large-scale printout of an iPhone and a cut-out window instead of the screen. Adam stood behind the screen 'window', while Coe simulated controlling his movements from in front of the 'screen'. With careful use of perspective illusion in the video recording from the laptop, the result was a good simulation of a real iPhone screen interaction. While not meant to be perfect, it certainly was sufficient to communicate clearly the key features and effect of the proposed feature and to facilitate discussion and valuable feedback from the viewers. See Figure 1.17. Search online for Elmo's Monster Maker MVP to see the video.

FIGURE 1.17   A screen shot from the prototype of Elmo's Monster Moves iPhone app, created in one hour.

One of the most downloaded children's apps, Elmo's Monster Moves designed by IDEO Play Lab, used a 'Wizard of Oz' illusion to demonstrate a working mobile interface that didn't exist. A simulation of the experience was created using some card props and members of the team to act out a use case, while a recording was made of just the screen area. Further information is provided in chapter 6.

execute

Develop,
deploy & scale

During the later stages of the Ideation and Validation phases it is likely that several concepts have been considered, evaluated, iterated and many discarded. A preferred concept has emerged and has been extended into a complete business proposal.

Many assumptions have been prototyped and tested. The innovator team is now ready to exploit the investment of time and money in research, reframing and prototyping. The concept and business proposal are prepared for execution of detailed development and market deployment. But, it is too early yet to go to full-scale production and launch; many questions remain about how the complete offering will perform in the heat of market battle.

Many decisions and more recent assumptions must be tested. What is the initial product configuration (MVP, minimum viable product) to be offered, which will provide learning about the market? What is the optimum business model that will sustainably deliver the innovation to the customer? Is the strategic alignment of the proposal with your own organisation clarified and are the strategic actions that are necessary for scaled deployment clear and prepared? Perhaps the proposed technology is found to enable some more new and potentially interesting features or instead provides some adverse limitations on what had been proposed; if so, how should you respond? Has the use-context evolved from what was researched months ago and should the concept and proposal be updated accordingly?

> The objective of Execute is not to earn revenue, but to prove revenue-earning capacity.

All of these questions have been considered earlier, throughout the innovation process, but now you must test them in real life. Also, you must test them with real, paying customer, albeit at a limited scale that allows rapid adjustment where necessary, and limits costs and possible reputational damage. The objective at this stage is not to earn revenue but to prove revenue-earning capacity. This is also a stage where substantial pressures are experienced by the innovator, both self-imposed and from the broader organisation and bosses. The innovator and others in the organisation are likely to be getting impatient. They have been excited by the prospects of the new proposal, and cannot wait to get it quickly into scale in the market.

As much as ever, process guidelines offered in chapter 7 are important for ensuring a successful launch.

In the next chapter, we start out on the ARRIVE process by introducing the techniques of Audit. Enjoy the journey!

BIBLIOGRAPHY

Alexander, Lameez and Van Knippenberg, Daan (2014). Teams in Pursuit of Radical Innovation: A Goal Orientation Perspective. *Academy of Management Review*, Vol. 39, No. 4, pp. 423–438.

Argyris, Chris (1983). Action Science and Intervention. *Journal of Applied Behavioral Sciences*, Vol. 19, No. 2, pp. 115–140.

Argyris, Chris (2004). *Reasons and Rationalizations – the Limits to Organizational Knowledge.* Oxford University Press.

Argyris, Chris and Schön, Donald (1978). *Organisational Learning: A Theory of Action Perspective.* Addison Wesley.

Baghai, Mehrdad; Coley, Steve; and White, David (2000). *The Alchemy of Growth.* Perseus Books.

Brown, Tim (2009). *Change by Design.* Harper Business.

Buchanan, Richard (1992). Wicked Problems in Design Thinking. *Design Issues*, Vol. 8, No. 2, pp. 5–21.

Cousins, Brad (2018). Design Thinking: Organizational Learning in VUCA Environments. *Academy of Strategic Management Journal*, Vol. 17, No. 2, pp. 1–18.

D'Aveni, Richard (1998). Waking Up to the New Era of Competition. *The Washington Quarterly*, Vol. 21, No. 1.

Dweck, C.S. (2007). *Mindset: The New Psychology of Success.* Robinson (GB).

Dweck, C.S. (1986). Motivational Processes Affecting Learning. *American Psychologist*, Vol. 41, No. 10, pp. 1040–1048.

Hassi, Lotta and Laakso Miko (2011). Conceptions of Design Thinking in the Design and Management Discourses, Proceedings of IASDR 2011, 4th World Conference on Design Research, 31 October–4 November, Delft, the Netherlands.

IDEO (2015). *The Field Guide to Human Centered Design*(1st edn). www.ideo.org.

Irish Government (2016). *Economic and Enterprise Impacts from Public Investment in R&D in Ireland. Irish Government Department of Jobs Enterprise and Innovation*, Dublin, Ireland.

Kelley, Tom and Kelley, David (2013). *Creative Confidence*. William Collins.

Kim, W. Chan and Mauborgne, Renée A. (2014). *Blue Ocean Strategy*. Harvard Business Review Press.

Lehman Darren R.; Chiu, Chi-yue; and Schaller, Mark (2004). Psychology and Culture. *Annual Review of Psychology*, Vol. 55, pp. 689–714.

Liedtka, Jeanne (2011). Learning to Use Design Thinking Tools for Successful Innovation. *Strategy & Leadership*, Vol. 39, No. 5, pp. 13–19.

Liedtka, Jeanne and Ogilvie, Tim (2011). *Designing for Growth*. Columbia Business School.

Moore, Geoffrey (2014). *Crossing the Chasm: Marketing and Selling Disruptive Products to Mainstream Customers* (3rd edn). Harper Collins.

Nagji, Bansi and Tuff, Geoff (2012, May). Managing your Innovation Portfolio. *Harvard Business Review*.

Norman, Don and Verganti, Roberto (2014). Incremental and Radical Innovation: Design Research vs. Technology and Meaning Change. *Design Issues*, Vol. 30, No. 1.

OECD and Eurostat (2005). *Oslo Manual – Guidelines for Collecting and Interpreting Innovation Data* (3rd edn). OECD Publishing.

PWC (2017). 2017 Global Innovation 1000 Study. Strategy& | PwC.

Rittel, Horst and Webber, Melvin (1973). Dilemmas in a General Theory of Planning. *Policy Sciences*, Vol. 4, pp. 155–169.

Tidd, Joe and Bessant, John (2009). *Managing Innovation* (4th edn), pp. 190–209. John Wiley & Sons.

Trott, Paul (2012). *Innovation Management and New Product Development* (5th edn), pp. 197–209. Pearson.

Verganti, Roberto (2016). *Overcrowded*. MIT Press.

CHAPTER 2

Audit

Design Innovation is a description of design thinking applied to business innovation. It combines deep human understanding, purposeful creativity and sound business rationale in order to create breakthrough innovations that succeed in the marketplace.

TODAY IS TOMORROW'S PLATFORM

The innovation team's starting point for innovation is often an incomplete vision or brief. The outline of intent is fuzzy, giving general direction rather than specific instructions. What does this outline intent imply? What does it entail? What is its scope? What are the big issues? Where do you start?

Start by improving your definition, scope and understanding. Research the market(s) where it may be targeted and the organisation where it will be developed and implemented. Look for opportunities with the greatest impact

FIGURE 2.1 Chapter 2 outline.

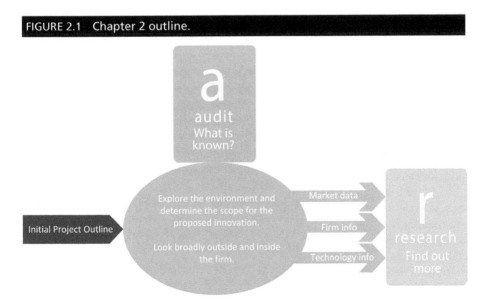

potential, and barriers with the greatest challenge. Look for societal and technological trends and key industry competitors. Sometimes, technology plays a central, driving role. On other occasions, it is merely supportive or perhaps enabling.

Your purpose in getting a better understanding of the status quo and relevant trends is to build a more solid project outline and scope, and to construct a platform for deeper human-centred research. Time spent here is investment rather than cost.

THE KICK-OFF

Suppose you are at a regular management meeting one Monday morning and your CEO, Jack, says to you:

> Joanna, I've been thinking over the discussion we had at the show last month. I agree absolutely with you; we need to up our innovation game in a big way and we can't wait any longer. I've had some discussions with Alex, our Board Chair, and he is coming around to the same conclusion. So, we've decided we'd better act straight away, and we think you're the person to lead this project! You've got the passion for this and we trust you. So, I'm giving you full responsibility to come up with our next major breakthrough, something to drive our growth over the next decade.
>
> We need something substantial, dramatic, radical; something that will shake the market and put us back on the map, leading the way like we used to do until a few years ago. I know you won't let us down and you know you'll have my full support. And, by the way, I've promised Alex we'll have something new for the market 12 months from now.

Your management colleagues look at you with strangely contorted features that express envy and relief at the same time. As you leave the meeting they offer you heartfelt congratulations and support, while confiding they are excitedly awaiting your presentation to the next meeting on your plans for the new project.

As you walk back to your office, your insides are full of mixed emotions. You are elated, but scared. You love the challenge but you are in awe of the uncertainty. It is a great career opportunity, but there are too many unknowns, and you can see chances to fail around every turn. Because of these risks, perhaps, your firm has been afraid or unable to undertake a major innovation effort up to now, despite aspirations being tentatively discussed from time to time over the past 12 months. But, thanks in part to your recent urging, senior management realises now that further inaction is even more unpalatable.

Where do you start? The ARRIVE framework is designed to deal with high stakes challenges like these.

Joanna's CEO got it right by recognising that her passion for innovation is an essential starting ingredient. An innovator must be passionate to succeed, of course. However, always and everywhere, passion intensity tends to decline over time, especially in the face of constant struggle against seemingly endless barriers and uncertainty. It is even more important to have the right mindset, a mindset for growth. We elaborated on growth and productive mindsets with their associated behaviours in chapter 1.

Joanna's first objective should be to work towards developing a collaborative coalition to support the project. The objective with a collaborative coalition is to have a channel of communication to the main body of the organisation and a forum where new ideas can be socialised and thrashed out. Operationally, this requires convening a multi-functional project team as early as possible. A multi-functional team may consist of representatives from Design, Marketing, Engineering, Manufacturing and Finance. Perhaps, other functions and business units may also be added, depending on the size and complexity of the organisation. Many organisations are already familiar with multi-functional project teams, for example in stage-gate processes, which focus usually on the later development stages of innovation projects.

Here, we advocate bringing together the team of disparate disciplines as early as possible, even at the research, ideation and concept-building stages, for two reasons. First, as the project develops, all disciplines can make useful contributions by bringing their various distinctive perspectives to user research, concept development, testing and selection. The second reason for early convening of a multi-functional team is even more important. Through early involvement, people gain a sense of ownership and anticipation. So, later on, when it comes to hard choices and perhaps difficult changes to ways of working, the potential resistance of team participants and their closer disciplinary colleagues, who will likely also be up to speed with the project, will be blunted.

The core team members and their closer disciplinary colleagues will be enthusiastic about the project's direction and prospects, hopefully. However, in larger organisations, a much larger cohort of staff and stakeholders will normally be untouched by the project until its launch is imminent. It is almost impossible to imagine a transformational market offering that does not also require a transformation in the organisation delivering it. Transformational change is the remit given to Joanna, so another aspect of the collaborative coalition-building is to socialise support more broadly throughout the firm. Joanna, directly and through her core team members, must cultivate familiarity, understanding and buy-in from key individuals throughout the organisation and influential champions at the highest executive levels.

Every organisation has its key influencers, strong change leaders that are well known and respected by staff employees and senior executives alike. By targeting key influencers to provide regular informal and semi-formal information updates about the project, and to seek advice in a spirit of open co-creation, future support for important decisions and widespread implementation will be more likely. Joanna already has the support of the CEO and Chair of the Board, which is essential. But often it is the next tier of senior management, if they are not aligned with the project, who can create subtle operational barriers to execution of a CEO's vision, where the busy CEO may not be able to sustain close monitoring of the project. In a McKinsey online article, 'Tapping the Power of Hidden Influencers', Duan, Sheeren and Weiss (2014) say that employee resistance is the most common reason for the failure of big organisational-change efforts, according to surveyed executives.

We have described in the last chapter how the ARRIVE platform for innovation has four pillars, relating to the user, market, technology and the firm (innovator organisation). The Audit phase seeks to understand all key driving parameters of three of these – *firm, market* and *technology*. Once a market area

FIGURE 2.2 The scope of Audit.

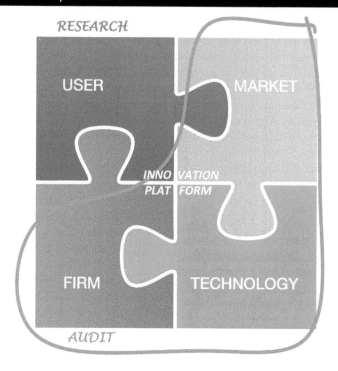

and use-cases have been identified, the jobs, needs, behaviours and aspirations of *users* will be explored and understood in depth using the techniques of the Research phase, which is the subject of chapter 3. See Figure 2.2.

What do you need in order to better understand the first three domains of firm, market and technology?

AUDIT THE FIRM

Understand the firm's strategic positioning and capability.

The firm's senior management may initially appear motivated. But, sometimes this is only skin-deep and you need to understand to what extent they and the broader firm are ready to handle the project you may present back to them, with recommendations for substantial organisational change, perhaps.

No organisations allow free rein for a new innovation project. There is always a strategic demand to retain connection with core business and existing strategy while being true to mission and vision. To establish an appropriate scope, ask the following questions:

- What are your firm's mission, vision and values? Be sure to articulate for yourself the mission-in-action as well as the espoused mission.
- What is its brand positioning?
- Does the firm have a clear, coherent, agreed and expressed innovation strategy aligned with a business development strategy? Not having a written strategy does not mean that there is no strategy operating in practice. If none exists, then you must deduce the effective strategy-in-action.

For one thing, you must consider the portfolio mix of innovation projects already alive in the firm, which will help you decide if there are already enough high-risk, radical projects on the books and if you should steer away from more of these. More often, the opposite is the case and there are too many incremental innovation projects focused on the short term, with a deficit of ambition to provide for medium- to long-term growth opportunities. The Innovation Ambition Matrix (Nagji and Tuff, 2012) illustrates nicely the parameters to be considered. See Figure 2.3.

Be sure to acquire a deep understanding of the underlying motivation for your project. What are you trying to achieve? There may not be a straightforward answer to this question.

Perhaps you are trying to defend or grow market share by improving your offering in a particular use context. This probably means improving the feature set of existing offerings or creating a broader set of offerings for that context. It may involve process enhancements or business model re-configuration.

Alternatively, you may be focused on driving growth through discovery of new market opportunities that match your firm's capabilities.

Maybe you have licence to be more adventurous and you have instinct or insight, based on trend analysis, which indicates there is opportunity for transformational creation of a new market-technology combination that has not yet existed.

Or, you may be seeking new market opportunities to exploit a key asset such as proprietary technology or fixed capital assets.

FIGURE 2.3   Innovation Ambition Matrix.

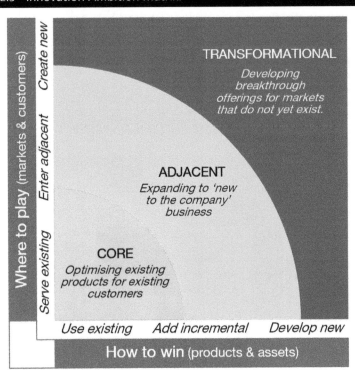

Source: Adapted from Nagji and Tuff, 2012.

There are many techniques in the realm of classical business strategic analysis and development to help with these issues, and their detailed treatment is beyond the scope of this book. Our purpose here is to alert you to the overarching issues that provide background, context and scoping for your project. It is important that you consider them thoroughly, so as to avoid the inevitable temptation to proceed too quickly to field-based research, discovery and development.

It is not enough to be aligned with the strategic direction and goals of the firm. Of course, practical limitations are also important, including the likely availability of budget and manpower to carry the project through to development and deployment in a reasonable timeframe.

Firm audit – summary

Innovation requires high energy and you have to be sensible where you expend it. If you fight the wrong battles too early you might not even get out of the starting blocks. An organisation will not change from conservative practices to bold moves overnight, so you must sensitise yourself to the organisation's brain and acceptable cadence of change, and build out from there. Ask yourself,

- What does the firm's current portfolio of products and services look like? What are its capability strengths? What argument and measurement is most likely to achieve buy-in?
- What is the expected meta outcome for the project? Is it more revenue, more margin, strategic market capture, capital asset utilisation (think of Amazon AWS), and so on?
- Where are the firm's constraints that are likely to impede innovation: finance, resource capabilities, mindset, key individuals, networks, etc.?

There is always choice with innovation possibilities. You must make informed choices and seek out the best available option rather than an unattainable perfection.

The right innovation is an ambitious proposal that gets deployed!

FIGURE 2.4   Firm Audit focus.

AUDIT THE MARKET

Where should you innovate?

In the scenario at the start of this chapter, it appears Joanna has a deceptively simple instruction – just innovate! Having clarified the firm's strategic motivation and project scope, more detailed questions arise concerning the markets, sectors, use-cases and territories where innovation opportunities might be found. As innovation leader, you can steer activities closer to your firm's existing business or further afield into less familiar areas. There is a balance to negotiate, continuously, and its resolution depends on context. Sometimes the area for focus is clear; it is where the challenge or opportunity has originated. In other circumstances, the only certainty may be that an existing market is saturated or dying and an alternative must be found, somewhere!

When a project is established to improve a particular use-case for a defined market, the focus and scope are well defined. This project will be quickly ready to begin acquiring the deeper understanding of the targeted user, which is the Research phase of chapter 3.

For many other projects the target market, use-case and user may be less well defined. For example, your current market may be eroding and you need to find new customers, but you are not sure where to look for them or what offering might be compelling. In this scenario, you will conduct a strategic analysis to identify new markets and possible use-case opportunities that identify potential users requiring deeper exploration.

What's the size of the prize?

A good way to get an overview of industry dynamics utilises the concept of *Profit Pools* (Gadiesh and Gilbert, 1998a), which seeks to identify the activities within an industry's value chain where margins are highest or, conversely, under threat, and thus opportunities or threats may be located. This requires considerable effort of analysis, but the insights can be worthwhile.

The total accessible market size (TAM) is an essential factor for all business efforts, no matter whether the markets are established, new or about-to-be-created. Even at early stages, the team must estimate market size using current best assumptions and perhaps, at least, in the form of a lower limit estimate. It is an important sanity check at all stages of the innovation project from research and conceptualisation through to development. Thus, also, budget checks and overviews should be reported at a high level at all times.

Business models

You must also understand the dominant business models that are used in the market. There have been transformative changes to business models in many industries – in fact, in most industries – over the last ten years, especially due to the flourishing digital economy and the internet of things. This is not just about revenue models and payment flows, although these have been at the heart of many transformations. A core resource of car sharing firm Uber, for example, is the ubiquitous citizen driver. Deployment of this resource is enabled by mobile digital technology, which also gives rise to a popular feature of the Uber service, cash-free transaction fulfilment, where payment is made via cloud account transfers. This latter feature has now also been introduced into the traditional taxi industry, and has proved so popular because it fixed a compelling problem of cash availability and cash handling for both types of users, passengers and drivers. Business models are changing rapidly in many industries. If they have not changed in your target market, this may be a threat and an opportunity.

Much of the above is the realm of traditional business and market development strategy, and later in this chapter we shall introduce you to our favourite frameworks and methods for analysing and capturing the most relevant information before you move forward with the project.

A practical issue in exploring new markets is that user-centred design innovation demands intensive study of users and other stakeholders to understand them, empathise with them and identify their key needs. The innovation team develops understanding and empathy through detailed research, involving dedicated effort and substantial time for each case. When the target user is not defined at the start of the project, the team must probe multiple areas, in order to establish the potential of each. Then, the team proceeds to deeper research of a shortlist of highest-potential opportunities.

Ask the experts

To learn more about a preferred market area to focus on, with its characteristics, pitfalls and opportunities, of course you should seek out the best expertise inside and outside the firm to inform and guide you. You would be crazy not to make the most of the rich resources on your doorstep or close by, although getting access to it may not be so straightforward. It is not just a download of data you want. By definition, an expert has been immersed in the target area for a long time, has encountered many challenges and difficulties, and understands the nuanced behaviours and predictable responses from all actors involved. This reflects the quote from the famous American physician and poet Oliver Wendell Holmes, Senior, from about 150 years ago: "The young man knows the rules, but the old man knows the exceptions".

Probably, every expert in the past will have had ideas or even more developed concepts that might have improved the experience and outcomes for the target users. Usually, such spontaneous thoughts are not immediately practicable, or even interpretable, due to the busy day job and other ongoing priorities in the firm, and they are archived indefinitely. It is extremely valuable to encourage experts to bring these ideas – old and new – to the surface and to engage them in critical and constructive discussion in support of your project.

There are many methods available for interrogating experts. The Delphi method is a long-standing structured communication technique that interactively interrogates a panel of experts about a future scenario (RAND Corporation, 2020). Recently, in his book *Overcrowded*, Roberto Verganti (2016), describes a method for drawing out innovation opportunities from experts using a structured process of critical dialogue progressing from *sole individual conceptualisation*, through *pair critical discussion* and then to larger groups. See the image in Figure 2.5. Around the same time, the book *Sprint* by Jake

FIGURE 2.5   Representation of innovation opportunity exploration through the method of criticism.

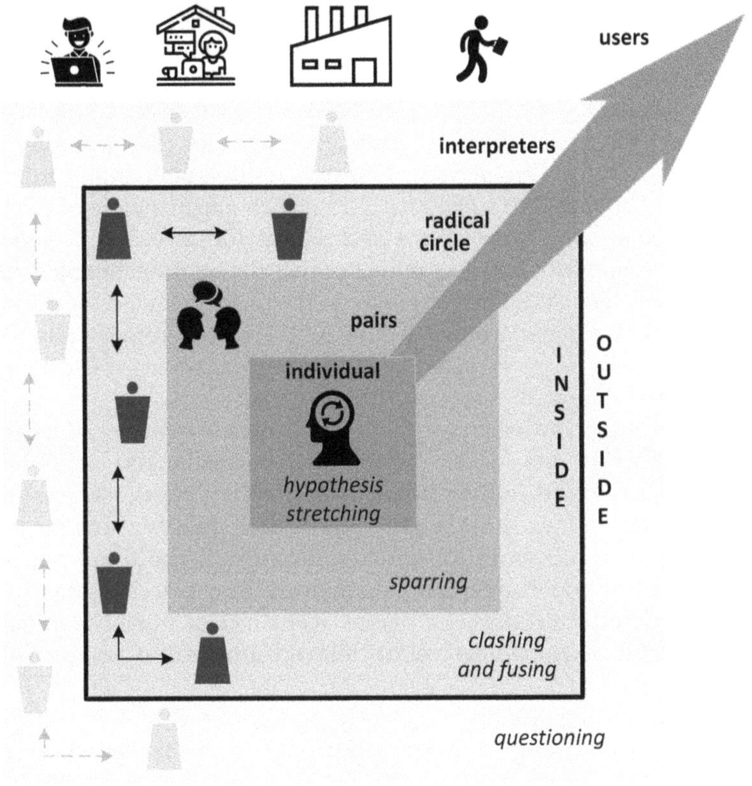

Source: Verganti, 2016.

Knapp (2015) and others from Google Ventures provides a method for interrogating experts to tease out the mapping of a target problem's focus.

Familiarity and superficiality – biases and assumptions

But, be careful. While you will draw on many sources of expertise, remember to treat the information you gather with appropriate scepticism. By appropriate, we do not mean you discard all information that is not 100 per cent proven. Instead, you recognise that truth is often submerged in information that is constructed by an interpreting narrative woven around extrapolations of narrow data. Such a narrative is probably based on intuition – good or bad – and usually has unconscious assumptions and biases implicit in it, not to mention your own cognitive biases that influence how you interpret that narrative (Liedtka, 2015). The sceptic with a growth mindset excels at uncovering and testing these assumptions and delights in finding a deeper and more solid interpretation to build upon. The experienced innovator recognises that uncovering key assumptions and reinterpreting a new narrative provides the golden nuggets of innovation opportunity.

You can see the need for caution by considering a typical scenario of a firm setting out to innovate in its 'home' sector, where it has been serving the market for decades and where it knows the customers, technology and competitive environment intimately. Often, firms working in familiar territory fail to notice gradual changes or evolutions in customer requirements or competitive offerings because they are too focused on the present solution types and their historical legacy that have brought success to date. Complacency as market leader can cause a firm to be an easy target for disruption by a new-entry offering. Clayton Christensen (1997) has written well on this topic in his seminal book, *The Innovator's Dilemma*. Often such a firm's view of the market is based on stale assumptions from a past that no longer exists. Business continues with a momentum that is waning because the relationship between firm and market, which existed harmoniously in the past, is now sadly out of tune. You need to fight against a blind acquiescence in this zombie state. A good motto is: familiarity is great, but unvalidated assumptions must be actively and constantly challenged.

The telephone industry provides an example where, 20 years ago, many landline telephone service suppliers were outflanked by mobile (cell phone) service providers in their home 'telephony' market. In typical disruptive fashion, cell phones initially offered inferior quality, causing landline providers to fail to recognise the threat, assuming it could not happen. As cell phone quality and convenience improved, and prices dropped, many landline telecommunications companies went out of business or suffered severely reduced margins because of delayed innovation responses to the disruptive threat.

Notwithstanding the above, clearly, ignorance is an even more dangerous threat. Attempting to innovate for users or a market where little is known is a recipe for disaster, a fact that is accepted in principle by most people. However, too often only inadequate remedy is attempted in the form of superficial desk research or poor, restricted surveys that seek to confirm existing prejudice, or high-level statistics with only sporadic interaction with that market's players. Additionally, it is not only the 'target' that must be researched. The firm's own capabilities and strategic ambitions must be queried and evaluated for compatibility and alignment, as we discussed above. The motto here is: research and evaluation must be comprehensive, thorough and appropriate, and all-important assumptions must be challenged and tested.

When Starbucks entered the Australian market in 2000, it seemed like a home run as Australian people undoubtedly love their coffee. But, Australians loved their coffee perhaps too much for Starbucks. They also treasure their local blends and the intimate atmosphere of the espresso bars around most corners, which have a central role in the community, in the same way as English people love their pubs. Australians, in their typically earthy manner, had little time for fancy Frappuccinos or milky macchiatos. Starbuck's failure to recognise these cultural and behavioural differences resulted in them pulling out of the Australian market eight years later with the closure of 60 stores and the loss of $143 million.

Each of these two extremes – home and away markets – and the myriad blended circumstances in between presents its own difficulties and dangers when setting off on an innovation journey.

FIGURE 2.6   Market audit focus.

Market audit – summary

Besides innovating to delight users, a successful project requires navigating the current industry structure to understand the market dynamics and ensure compatibility with it. In particular, the innovation team must understand relevant trends, and how they are likely to develop over the coming years.

- Which markets offer the best potential for satisfying the innovation objective? What are the relevant use-cases and trends?
- What is the competitive landscape? Who are the competitors, direct and indirect, and what is their likely competitive response to your new innovative offering?
- What are the dominant business models, established and new? How might these change, especially with the onward surge of the digital economy?
- What are expected revenues, margins, resource velocities, investment returns and financial health of the main competitors? Where are the profit pools within the industry?

AUDIT THE TECHNOLOGY

What special know-how will be needed to implement the solution?

Non-technology-driven businesses

For markets that are not technology-intensive it may be sufficient to give Technology audit a cursory evaluation and to cover some of the topics under 'Firm' and 'Market' audits as we indicate below. However, having said that, understanding of technology requirements for an innovation has become more important for all industries in recent years because of the ubiquity of technology based solutions in the digital economy.

Some markets are technology agnostic, i.e. the customers do not care which technology is used so long as the product and experience are as expected. Others, like gaming and telecommunications for example, are technology fetishist where specific or latest technology is demanded.

You do not always need radical new technology to create a breakthrough innovation, but usually you need a novel or at least a competent application of existing or state-of-the-art technology. Development of new technologies carries huge uncertainty from many perspectives like performance, cost, and time to deployment. However, even with established technology, developing products still requires substantial effort, resources and competence.

Of course, not all technology is 'digital' or electronic. By technology we mean any special, codified know-how, and it covers all of scientific knowledge and beyond, from pharmaceuticals to oil exploration, and from rocket science to bag-less vacuum cleaners.

Technology management is complex

In technology intensive firms, technology management is complex and covers a broad range of topics, such as technological trajectories, technology life cycles, technology roadmapping, a firm's core competencies and technology strategy. Technology management is so complex and important that it is given a special status in most organisations.

To illustrate the special treatment of technology research and development, note that it generally incurs heavy expenditure, even up to 20 per cent of revenues in some industries. It is often reported separately at a high level (as R&D) in firms and even national accounts. It tends to be oriented towards building a platform or basic capability rather than dedicated to a specific product or project. It is often partly done outside the firm by collaborating with technology research firms or universities. And, it tends to take place over longer time scales. For these reasons and the fact that it is extensively covered in many other sources, we do not dwell on technology research or management in this book. We provide some useful references for further reading on technology research and management at the end of this chapter (Schilling, 2010; Cohen and Levinthal, 1990; Cooper, 2017; Cetindamar et al., 2010). Nevertheless, there are a few special points that an innovation team should seek to understand in preparing a realistic vision for a new innovation within a real business, as we outline below.

Absorptive capacity

It is not easy for a firm, nor for an individual, to acquire a new technology where it has had no experience previously, even if it is a technology that is well established in the wider world. Cohen and Levinthal (1990), in their seminal paper, describe the importance of absorptive capacity in a firm to its "ability … to recognise the value of new, external information, assimilate it and apply it to commercial ends". This prompts some questions for the audit in preparation for the innovation project, and the answers feed into the strategic setting for your planned innovation, especially for technology intensive businesses.

Trigger questions

Some useful questions for your technology audit are below.

Market (this may be included in the 'Market' audit, for non-technology firms)

- Is your target market strongly defined by technology? As we have said, some markets are defined by strong technology-related performance and user awareness, while others are agnostic.
- Is there a significant technology disruption underway or likely to happen in this market in the next 3–5 years?

Firm (this may be included in the 'Firm' audit, for non-technology firms)

- What technological competences does your firm have, if any?
- What is the technological absorptive capacity of your firm? (This is closely related to the preceding question.)

(For technology intensive firms)

- Does your firm have a technology management and development strategy? If it is not explicit and clear, then you need to understand what the implicit strategy-in-action may be.
- What are the technology trajectories and trends in the market sector that you propose to serve?
- What are the typical times-to-market associated with new products using technology similar to that proposed for this project? And, what are the typical lifecycles of technology generations.

If your firm does not have a strong or the right technology competence, it does not immediately mean you should abandon the project but it may alter the business model by which you can exploit the market opportunity you have identified.

Technology audit – summary

FIGURE 2.7   Technology Audit focus.

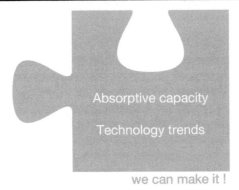

Understand your firm and your target market:

- What is the technology management strategy of your firm?
- What are the firm's core technological competences and absorptive capacity?
- What are the dominant technologies or special capabilities used in the marketplace?
- Are there likely to be technology disruptions in the market over the next 3–5 years?

THE AUDIT PHASE IS ESSENTIAL

Getting yourself well informed about all the parameters we have described in this Audit phase above can be a challenging task. Nevertheless, please resist any temptations to skip through it superficially! When a person or team is tasked with innovating, sometimes these early stage investigations appear to be a distraction, a time-wasting, frustrating delay before getting to the meat of the innovation. We all wish to be out in the field creating great new ideas and showing them off to the client. We like to impress our senior management and shareholders with the early fruits of our creative ingenuity. The frustration with being held back from the action, straining at the leash, is understandable.

However, research clearly shows that this phase at the fuzzy front-end is a crucial determinant of success in innovation projects. The authors have consistently found that time spent understanding the problem space is an investment, rather than a cost. Proper attention and time given to the methods and considerations of the Audit phase outlined in this section are richly rewarded. For novices in a new application area, it is clearly essential. Even for domain experts, it is important to step back, reflect, take a bird's eye view of the domain and beyond to see how the structure or the dynamic of the domain may have changed over time. These days, more change is always imminent, for good or ill. It is rare to find a firm that has sustained a perfect balance. On the one hand, there is expertise won through experience and success in continuing operations; on the other hand, everyone needs to take time to reflect and gain deeper understanding of new trends and influences and their potential strategic impact on the business's future. Before setting out on a long and expensive innovation journey it is essential to fully understand the present so that it may be a solid platform for the new future that you are creating.

METHODS IN THIS CHAPTER

Many of the methods used in the Audit phase and described in this chapter are those of classical strategic analysis, and facilitate both internal (to the firm) analysis and external industry, market and global analysis.

There are well-known and excellent methods in general use and we have selected a small set of our favourites to present here, methods we consider best to feed into the later stages of the ARRIVE process. You should note that our treatment of these classical strategy frameworks is cursory, but we hope enough to encourage thought, reflection and better understanding of the innovation environment which will inform the detailed research and conceptualisation of later stages in the project. If you are new to these methods, we encourage you to read and study further.

To understand the broader ecosystem and the industry environment, we use the well-known traditional frameworks:

- PESTEL
- Porter's Five Forces

To assist preliminary identification of possible sub-markets, users and use-cases, where the target markets and users are as yet undetermined or unclear, we use:

- Profit Pools
- Customers and Non-customers (Blue Ocean)
- Process Chain Mapping

The methods and frameworks to understand the firm environment that we illustrate are:

- Statements of Mission | Vision | Values
- Firm's Strategic Innovation Profile

Finally, at the project level, we recommend clarifying your initial understanding of the project brief which provides initial focus and sharing with the team:

- Initial Project Outline

OUTPUTS AND TEMPLATES

At the end of your Audit phase, you have a set of collateral assets, some in the form of completed templates that act as a project guide set for later stages. Outputs from this stage are:

- **Initial Project Outline:** Every project must have direction and the *Initial Project Outline* provides this. However, innovation is a special learning process and it should not be overly constrained too early. Discovery of radical new ways to add value usually happens organically, when the team has freedom to explore new learning opportunities, which are not foreseen or foreseeable from the start. Hence we emphasise the word *initial* in the *Initial Project Outline*. It starts off as a guide but should be considered to be malleable and amendable in response to new learning uncovered throughout the project. The *Initial Project Outline* also incorporates the primary motivation for the project with identified possible areas of interest and other scoping parameters. See *our Initial Project Outline* method later in this chapter.
- **Firm's Strategic Innovation Profile:** Innovation teams, sometimes, may be carried away with the excitement of learning, creating, innovating. It can be easy to forget, for a while, that an innovation will only exist when it is deployed. As we have said, the firm must be capable and willing to allocate resources and put in place effective processes, strategic planning and execution that are necessary to ensure deployment. We have found that this is a regular crunch point at a late stage in too many projects, and has the consequence of much frustration, wasted effort and sometimes major project reworking or cancellation. Our template *Firm Strategic*

Profile is a visible reminder throughout all stages of the project that the outcome must be suitable for the firm. See the *Firm's Strategic Innovation Profile* method later in this chapter.

- **Environmental information:** The project team must understand well the market and technological ecosystem into which the successful innovation will fit. This involves the basic demographic, consumption and competitive data, in addition to social and technological influences and trends. We provide various methods to assist building this understanding, later in the chapter.

Audit | methods

a
audit
What is known?

r research
Find out more

r reframe
Change Perspective

i ideate
New Ideas & Concepts

v validate
Test with users

e execute
Develop, deploy & scale

1

2

PESTEL
PORTERS FIVE FORCES

PROFIT POOLS
CUSTOMERS AND NON-CUSTOMERS
PROCESS CHAIN MAPPING

3

4

MISSION | VISION | VALUES
PORTERS FIVE FORCES

INITIAL PROJECT OUTLINE

To understand the broader ecosystem and the general industry environment, we use the well-known traditional frameworks:

- PESTEL
- Porter's Five Forces

To assist preliminary identification of possible sub-markets, users and use-cases, where the target markets and users are as yet undetermined or unclear, we use:

- Profit Pools
- Customers and Noncustomers (Blue Ocean)
- Process Chain Mapping

The methods and frameworks to understand the firm environment that we illustrate here are:

- Statements of Mission | Vision | Values
- Firm's Strategic Innovation Profile

PESTEL ANALYSIS method 1/4

"No man is an island, entire of itself;
every man is a piece of the continent,
a part of the main."

John Donne, 1624

Nor is a firm an island! Nor an innovation! While it is natural to be most familiar with and to think first about the firm and its industry, with their opportunities, challenges and dynamics, it is also important to understand the broader ecosystem, in which the business and industry operate. The PESTEL acronym provides a useful framework with which to interrogate this broader environment.

Some of the PESTEL factors may have immediate relevance. Others may have more peripheral influence at the moment. You need to look out for trends or impending events that might influence your project, perhaps to a dramatic extent.

Step by step

1. **Get started, then keep at it**
   The PESTEL analysis is rarely a once-off, static compilation. It will be updated regularly throughout the project as new information comes to hand or as new situations unfold in the environment.

   Start by convening your team for a 90-minute brainstorming and analysis session.

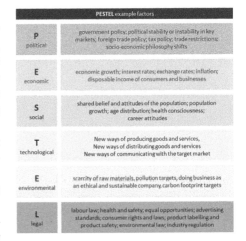

2. **Identify issues relevant to the project**
   Use Post-its for freeform brainstorming, initially, to identify any issue, event or trend in the broader environment that may influence your project (20 minutes). Look at the present and the foreseeable future. One issue per Post-it.

   Spend ten minutes sorting the Post-its into the most relevant category of PESTEL. Don't worry about exact placement, e.g. between P and L, etc. Use whichever seems a most natural fit. You end up with six categories and multiple issues in each.

3. **Repeat, with more focus**
   Repeat the brainstorm, now focusing on each of the PESTEL categories in turn, to add new issues to your list. Take five minutes per category, 30 in total.

4. **Rate the issues in each list**

Spend 30 mins to rationalise the lists and assign a rating (*low, moderate, high, very high*), for both 'uncertainty' and 'importance', to each issue. See the template on the next page.

5. **Locate: ACT | WATCH | SCHEDULE | IGNORE**

Map the issues (Post-its) onto a map as shown in the example page of this method.

You now have a guide to the actions to be taken with each issue you have identified.

6. **Revise regularly**

Keep updating and referring to your PESTEL issues list and Action Map as you learn more going through the project, from industry experts, industry reports and your own increasing experience.

PESTEL ANALYSIS template   2/4

FIGURE 2.8 PESTEL analysis template.

PESTEL analysis

date	name			Description	* Score 1-4 1 = low 2 = moderate 3 = high 4 = very high	Importance*	Uncertainty*
Rev.	Code						
P political							
E economic							
S social							
T technological							
E environmental							
L legal							

PESTEL ANALYSIS   example

FIGURE 2.9   PESTEL analysis for an Irish Health Insurer.

date	Feb 2019	name	ABC Health Insurer	PESTEL analysis	* Score 1-4 1 = low 2 = moderate 3 = high 4 = very high	Importance*	Uncertainty*
Rev.	1	Code		Description			
P political		P1	Growing political pressure to remove private healthcare activities from public hospitals			3	3
		P2	Conditions are still uncertain for and after upcoming BREXIT			4	4
		P3	Risk Equalisation payment transfer mechanism continues to be an active public discussion topic.			1	2
E economic		Ec1	Rising medical inflation globally			2	1
		Ec2	Increasing competition from European and US suppliers			3	1
		Ec3	Former monopoly still dominant in the market			2	1
S social		S1	Ageing population. Over 65s projected to rise from 13% (2013) to 25% (2050)			3	1
		S2	Personal behaviour trends: smoking down; alcohol consumption up; obesity up.			2	2
		S3	Falling birth rates. Older first-time mothers.			2	1
T technological		T1	Continuous advances in medical technology and associated costs			3	1
		T2	Big data through fitness wearables			2	3
		T3	InsureTech emerging globally, enabled by new technology, e.g. Blockchain, IoT, cashless payments			3	2
E environmental		En1	Greening economy with reduction in pollutants and associated adverse health effects			2	2
		En2	Paperless transactions			3	1
		En3	--				
L legal		L1	Increasingly strict regulation			2	1
		L2	Lifetime community rating system changes			3	3
		L3	--				

FIGURE 2.10   PESTEL Issues Action Map.

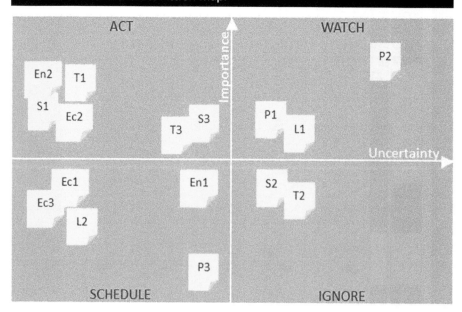

Q My team is made up of designers, marketers, engineers. How can we be expected to know all geo-political and socio-cultural issues that are relevant?

A As we say often, simply doing the exercise and sensitising yourself to the types of issues involved also has its own value.
Probably, you will catch over 80 per cent of important issues. But, don't think it stops after one meeting. You should continue to update the issues list and the Action Map as you gain more information throughout the project. After your first meeting, bring the map to some senior execs in the firm and further afield around the industry to also get their input.

Q I have not heard of the PESTEL Action Map before. Why use it?

A This book expounds design thinking and we believe in the power of visualisation and action-prioritisation. Some PESTEL exercises produce a long list of issues, where it is hard to see the wood for the trees.
We created the Action Map to provide an at-a-glance display of the important issues, and guidance for actions to be taken.

Q It is hard sometimes to agree the category where an issue belongs. How can we resolve differences of opinion between team members?

A We are tempted to say you should toss a coin! Often it doesn't matter. You may count regulation as a political or legal issue, or you may say that labour rates are an economic or political or social issue or all three!
The PESTEL categories are simply convenient buckets with which issues may be identified and captured. The placement of the issue on the Action Map will be the same regardless of the 'bucket' that was used to transport it there.

Q How many issues would you expect to raise?

A How long is a piece of string? By 'issue', we mean a trend or a recent or impending event that may influence your project. It is hard to imagine that there won't be at least one or two of these in each category, maybe many more. But don't sweat it. Give the exercise the honest 90 minutes to start. Then canvas others who should know, and revisit it regularly. It is unlikely you will miss anything really significant.

The Five Forces is a framework for understanding the competitive forces that are at work in an industry, and which drive the way economic value is divided among industry actors. First described by Michael Porter in his classic 1979 *Harvard Business Review* article, subsequently updated (Porter, 2008), Porter's insights started a revolution in the field of strategy and continue to shape business practice and academic thinking today.

A Five Forces analysis can help companies assess industry attractiveness, how trends will affect industry competition, which industries a company should compete in, and how the company can position itself for success.

The key industry structure concepts are the following:

1. Every industry is different, but the underlying drivers of profitability are the same in every industry.
2. The Five Forces determine the competitive structure of an industry, and its profitability. Industry structure, and a company's relative position within the industry, are the two basic drivers of company profitability.
3. Analysing the Five Forces helps companies anticipate shifts in competition, shape how industry structure evolves, and find better strategic positions within the industry.

Conducting a rigorous, quantitative analysis of all five forces is a strategy specialist's domain, and requires time and access to plentiful relevant data. If you do not have these available, still you should be sure to consider, as much as possible, this important perspective on the industry you are targeting for your innovation project. We give you guidelines for an initial consideration here.

Step by step

1. **Define your industry**
   What is the industry where your firm is active and what is its underlying competitive structure? This information is essential in order to identify where defensive reaction is required or, alternatively, where opportunities arise to be proactively exploited. Define the industry along two scoping dimensions: product and geography.
2. **Identify participants**
   Identify participants and segment them into groups, as appropriate. Who are the:

   Buyers and buyer groups?

   Suppliers and supplier groups?

   Competitors?

   Substitutors (and their products)?

   Potential entrants?

3. **What is the underlying strength of each competitive force?**
   Identify the main drivers for each competitive force and estimate the forces' relative strengths. Be clear as to why these relative strengths (and weaknesses) are as you have represented.

4. **Identify likely changes**
   Analyse recent and likely future changes or trends in each force, positive and negative.

5. **Map and identify**
   Map the findings on the template of the following page (which probably you will need to extend). Identify: the top two areas of threat to your firm's existing business, and the top two areas of opportunity.

6. **Revise regularly**
   Keep updating and referring to your Five Forces map as you learn more going through the project, from industry experts, industry reports and your own growing experience.

FIGURE 2.11   Five Forces of Competition template.

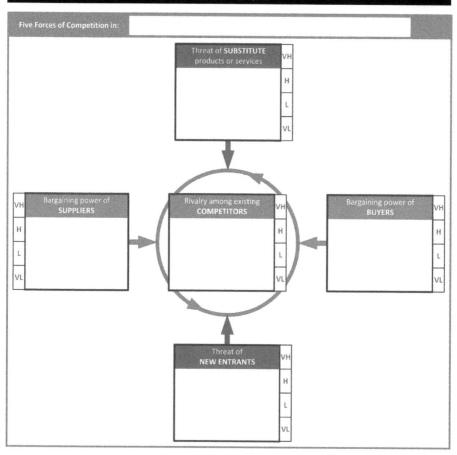

FIVE FORCES OF COMPETITION   example   3/4

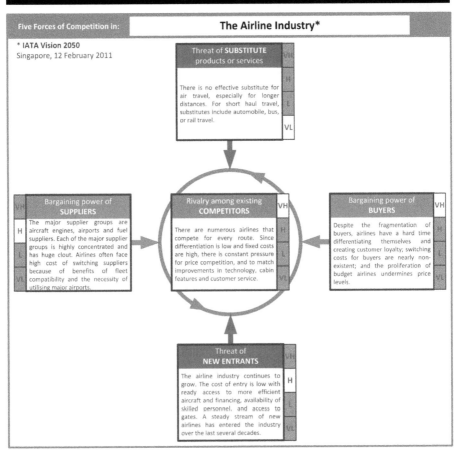

FIGURE 2.12   Five Forces of Competition for the Airline Industry 2011.

Five Forces of Competition in: **The Airline Industry***

* IATA Vision 2050
Singapore, 12 February 2011

Threat of **SUBSTITUTE** products or services

There is no effective substitute for air travel, especially for longer distances. For short haul travel, substitutes include automobile, bus, or rail travel.

Bargaining power of **SUPPLIERS**

The major supplier groups are aircraft engines, airports and fuel suppliers. Each of the major supplier groups is highly concentrated and has huge clout. Airlines often face high cost of switching suppliers because of benefits of fleet compatibility and the necessity of utilising major airports.

Rivalry among existing **COMPETITORS**

There are numerous airlines that compete for every route. Since differentiation is low and fixed costs are high, there is constant pressure for price competition, and to match improvements in technology, cabin features and customer service.

Bargaining power of **BUYERS**

Despite the fragmentation of buyers, airlines have a hard time differentiating themselves and creating customer loyalty; switching costs for buyers are nearly non-existent; and the proliferation of budget airlines undermines price levels.

Threat of **NEW ENTRANTS**

The airline industry continues to grow. The cost of entry is low with ready access to more efficient aircraft and financing, availability of skilled personnel, and access to gates. A steady stream of new airlines has entered the industry over the last several decades.

Q Does the Five Forces framework just involve a listing of competitors in each category?

A It is much more than that. For a comprehensive strategic analysis of your industry using the Five Forces framework, we recommend you read some of the many detailed publications on the topic. We have provided references to some of these on the introductory page to this method.

It is almost worthless to complete the Five Forces template with just a free brainstorming, and then to leave it as a simple listing. You get most value by considering each category in detail, identifying factors that determine its strength or weakness, e.g. switching costs, concentration of buyers or sellers, regulation. Then, deduce which actors these factors favour or hinder, and their profitability impact. This is done most thoroughly by quantitative analysis of industry and firm data.

As we said earlier, if you don't have the time, expertise or data to conduct a full analysis, there is still value in making qualitative assessments, and flagging some areas for more detailed analysis later.

Q Are we looking at the forces of competition now or in the future when our innovation will be deployed?

A Both. You can't predict the future so you need a good start with the present. Once you have clearly characterised the present competitive situation, you must then dig deeper and identify trends that might portend the future (step 4 in our method steps).

Look at recent changes and see if they are indicative of a trend that may carry on into the future. Then look at other anticipated environmental changes (regulatory, technological, convergence, etc.) to see whether they might have an impact.

Q What timescale should we consider for our trend analysis?

A That depends on your industry. Some industries have shorter reaction times than others. However, be careful not to confuse cyclical or seasonal variations with genuine underlying structural changes of the industry.

PROFIT POOLS method

Follow the money …

Understanding the market also means knowing the financial indicators that characterise its performance and its main players. Are these stable or are they undergoing (or likely to undergo) substantial change? You should be familiar with typical revenues, margins, profitability, resource velocities, investment returns and the financial health of the main players throughout the industry value chain. A key diagnostic for many industries is to figure out where the profit pools are within the industry, a technique pioneered by consultants Bain & Company (Gadiesh and Gilbert, 1998b). You can explore this by mapping out an industry value chain and the profits taken from each stage.

This can be a laborious exercise, if done properly, and may need specialist industry strategist's help. But, the rewards can be great insight into where opportunities lie within your industry.

Take the car and home insurance market for example, where, in recent years, policy prices (premiums) have been going up but insurance companies still find it hard not to lose money because of more frequent random events like flooding or other extreme weather events. If you were to do a profit pool analysis, you would find that the most profitable part of the market is the category of aggregators (Compare the Market or Go Compare, etc.). Aggregators take a flat percentage of every policy sold but they don't bear the risk of a payout in the event of accident or claim. Studying a value chain and profit pool analysis often reveals some surprising potential sources of growth.

Step by step

1. **Define the pool**
   Take a broad view of the value chain of your industry. Don't disaggregate activities more than is necessary. Determine which value-chain activities influence your ability to generate profits now and in the future. Create a list of all value-chain activities in your profit pool, in sequential order.

2. **Determine the size of the pool**
   Develop a baseline estimate of the cumulative profits generated by all profit pool activities. Seek a rough estimate, but as accurate as possible within a timeframe. Take the easiest analytical routes available; go where the data are. Try to take two alternative perspectives, e.g. company level and product level. Express the resulting estimate as a probable range.

3. **Determine the distribution of profits**
   This is the key challenge. Develop estimates of the profit generated by each activity. If relevant company data aren't available, use proxies such as product-level or channel-level sales. Produce separate estimates of the profits for each value-chain activity.

4. Reconcile the estimates

Compare the outputs of steps 2 and 3 and reconcile the numbers. If they don't add up, check all assumptions and calculations to resolve all inconsistencies. Produce the final estimate of the profits for the total pool and all activities within the pool.

5. Map the profit pool

Plot the value chain activities' profitability against revenues, to see the relative profits being generated by each activity. As well, by comparing a profit pool map with a map from an earlier time, you can spot trends in profit distribution.

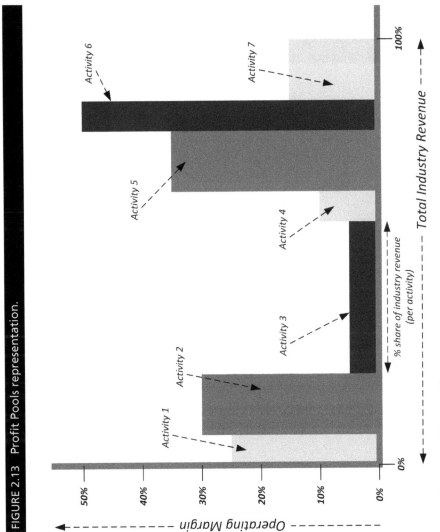

FIGURE 2.13 Profit Pools representation.

Source: Gadiesh and Gilber., 1998b.

FIGURE 2.14   Global Healthcare profit pool 2010.

Global healthcare profit pool (2010)

EBIT margin   Total=$520B

30%

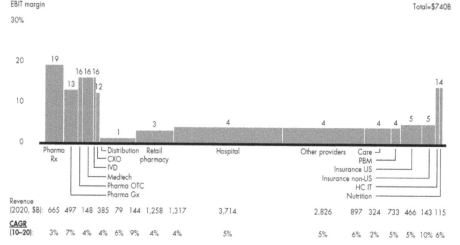

Revenue
(2010, $B): 495 230 100 260 44 62 850 890 2,280 1,733 501 266 450 260 55 64

FIGURE 2.15   Global Healthcare profit pool 2020.

Global healthcare profit pool (2020)

EBIT margin   Total=$740B

30%

Revenue
(2020, $B): 665 497 148 385 79 144 1,258 1,317   3,714   2,826   897 324 733 466 143 115

**CAGR
(10–20):**   3% 7% 4% 4% 6% 9% 4% 4%   5%   5%   6% 2% 5% 5% 10% 6%

Note: Insurance US reflects total health plan premiums; CXO represents contract research, manufacturing and sales organizations
Sources: IMS; Datamonitor; Business Insights; Freedonia: annual reports; analyst reports; CMS; OECD; Bain analysis

In June 2012 Bain & Company published an Insight Brief showing profit pools in the global healthcare industry and how they were expected to change in the period from 2010 to 2020. See Figures 2.14 and 2.15, where it is interesting to see how branded pharmaceuticals show low margins growth of 1 per cent for the period while generics show a much healthier 7 per cent. Overall, most of the growth in margins was expected to come from increased volume in the delivery of care. Clearly, this is very useful information for a business looking for growth opportunities in the healthcare market around 2010.

Source: Used with permission of Bain & Company. https://www.bain.com/insights/healthcare-2020/.

Q This seems daunting. How can I get accurate data for all the players in my industry?

A You won't get accurate data for each player. But, you can get data for the really big players, especially if they are public firms. An 80/20 rule often applies. If you have good data for the really big actors in an industry, you may extrapolate margins etc. to the remaining 20 per cent of smaller actors.

You should make use of industry reports, product sales information and all other relevant available data.

Q How many activities should there be in an industry value chain?

A The short answer: as many as makes sense! An activity that has distinctive function, structure or business model and has substantial scale is probably worth nominating as a separate activity. However, try not to have too many activities, with too fine distinctions, which will result in difficulty parsing the available data to match them.

Q What is the purpose of constructing a profit pool analysis?

A Ideally, you might identify an emergent activity within the industry that is showing really high margins. Certainly, that would then be well worth exploring further to see how you might increase your firm's involvement. Alternatively, you might identify an activity where margins are coming under pressure from other changes.

At any rate, you will certainly benefit from understanding the structural and margin dynamics of your industry more deeply.

Q I'm not a business or market analyst. Is it worth it?

A Established industries are evolving rapidly, due to business model and technology changes and new entrants from globalisation. It's risky to ignore what may be already happening but is obscured by the complexity.

This is a method for clarifying industry complexity. If you are contemplating a strategic move within your industry, it's a perspective that is risky to ignore.

Porter's Five Forces framework illuminates the competitive structure of a market and the competitive positioning of its various actors. The key objective is to determine which part of that market offers the greatest opportunity and which strategy is appropriate to win there.

Blue Ocean Strategy (Kim and Mauborgne, 2014, 2017) addresses the segmentation problem from a contrasting perspective. Rather than assuming the boundaries, actors and competitive structure of a market are fixed and hence define a red ocean of intense competition, the Blue Ocean approach seeks out new uncontested space (blue oceans) that make the competitive structure of the existing market irrelevant. It explicitly seeks to move the goalposts of competition in the industry.

One step in the Blue Ocean process is to identify those people who are not presently customers of your industry's offerings. The Blue Ocean authors have created a framework of Three Tiers of Noncustomers, which is very useful for exploring where new demand might be found.

FIGURE 2.16   Three Tiers of Noncustomers.

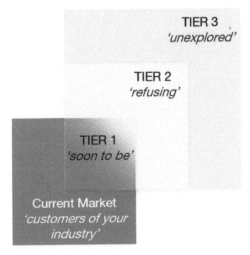

Source: Blue Ocean Strategy. Kim and Mauborgne, 2005.

Step by step

1. Convene your team for a four-hour brainstorming and analysis session.
2. **Who are your (industry's) existing customers?**
   Define the major segments. Identify three reasons that customers are attracted to use the offerings from you and your industry competitors. Also, where do existing customers have pain points tending to push them away? Stay at a high level; avoid fine detail.

3. **Who might be Blue Ocean noncustomers?**

   Allocate 15 minutes to each of the three tiers of noncustomers, in turn. Start with Tier 1. For each tier, have the team write on Post-its the names of customer groups they identify in that tier. Encourage discussion. Place all Post-its on a section of whiteboard labelled for that tier.

4. **Review**

   Spend a further 30 minutes to review all customer groups and their allocated tiers. Rationalise any duplication of groups and remove any 'undecided'. You now have a whiteboard with three 'tier' sections and a number of customer groups per tier.

5. **What's the size of the prize?**

   Ask your team to spend a half-hour on-line, individually, to estimate the potential demand from the various customer groups. Settle for high-level estimates; avoid discussion of fine detail. Discuss and agree high-level estimates.

6. **Revise regularly**

   Choose the top five customer groups by demand size. Discuss each in turn. Agree the tier to which each belongs and assign up to three 'push' and 'pull' factors. What are the factors that might attract each customer group to your industry's present offering and what might draw the customer group away from your present offering? Complete the Customers and Noncustomers template overleaf. You now have a rich shortlist for deeper opportunity investigation.

CUSTOMERS AND NONCUSTOMERS template   2/4

Three tiers of noncustomers

Soon-to-be noncustomers
... patronise your industry because they have to, not because they want to. They are wavering at the edge.

Refusing noncustomers
... have consciously thought about, but rejected, using your industry's offering. It's not attractive to them.

Unexplored noncustomers
... have never been thought of as potential customers, nor targeted by any of the industry's players.

FIGURE 2.17   Customers and Noncustomers template.

INDUSTRY CATEGORY:

	top 3 pull factors	CUSTOMERS & NON-CUSTOMERS	top 3 push factors

EXISTING CUSTOMERS

pull side: 1, 2, 3
push side: 1, 2, 3

Soon-to-be | Refusing — *select one* ⤢ **Unexplored**
Customer group:
Demand size: (M€)
pull: 1, 2, 3   push: 1, 2, 3

Soon-to-be | Refusing — *select one* ⤢ **Unexplored**
Customer group:
Demand size: (M€)
pull: 1, 2, 3   push: 1, 2, 3

Soon-to-be | Refusing — *select one* ⤢ **Unexplored**
Customer group:
Demand size: (M€)
pull: 1, 2, 3   push: 1, 2, 3

Soon-to-be | Refusing — *select one* ⤢ **Unexplored**
Customer group:
Demand size: (M€)
pull: 1, 2, 3   push: 1, 2, 3

PULL to your offering

PUSH from your offering

CUSTOMERS AND NONCUSTOMERS example 3/4

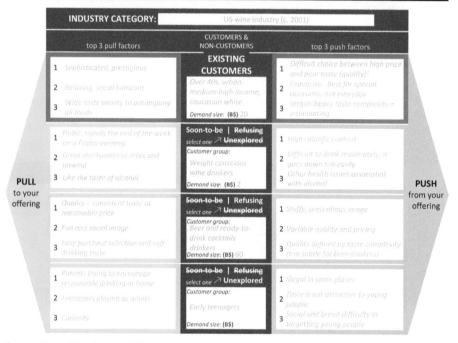

FIGURE 2.18 Customers and Noncustomers example.

Source: Kim and Mauborgne, 2005.

Q Surely, there must be plenty of people who are customers of my competitors, but not of my own firm. Why do you talk about noncustomers of my *industry*?

A Traditional strategic approaches concentrate within the current industry boundaries and thus must be concerned with taking customers from competitors. The beauty of Blue Ocean Strategy is that it takes an alternative approach and emphasises looking outside the industry, where there may be unexplored opportunities that have substantial demand and are easier to exploit.

Q Is this the full Blue Ocean Strategy?

A No. Blue Ocean Strategy is a rich and well-rounded approach to strategic growth development. The Three Tiers of Noncustomers is just one framework within Blue Ocean Strategy, and we like it! That is, particularly, because it is nicely aligned with our own Design Innovation approach to strategic innovation.

Q Tier three noncustomers are simply everyone else not included in existing customers and the other two tiers. Is that right?

A No! That is far too broad an interpretation. 'Unexplored' refers to noncustomer people or organisations whose needs and jobs-to-be-done are such that they *should* benefit from the core offering of your industry. However, this has not been recognised by your industry, which has made itself somehow unattractive, unimaginable or unfeasible.

Q Is it not important to do this exercise properly and in detail? How can this be achieved in just a four-hour session?

A As we often say, you can save time and effort by not delving precipitously into too much detail. The time for fine detail is when you are refining and finessing a chosen strategy concept. At earlier stages, it is important to concentrate on the big picture and a lot can be achieved in a half-day, with focus. Additionally, this exercise will not be the sole determinant of your whole innovation strategy. But, it will make a great contribution.

This is a method to uncover possible opportunity areas for innovation in a given context.

Use Process Chain Mapping when you are doing a high-level review of a use-case context or process and you want to identify where opportunities might exist for adding value to the supply ecosystem of an existing product or service.

It helps you to address the questions:

- How can an existing process be improved?
- Are there simpler ways to achieve the same or better outcome?
- How might we more efficiently or effectively deliver a better service to the end user?

Process Chain Mapping is a simple method for exploring and deconstructing a known product delivery system in order to query each component and find what might be eliminated or reduced. It is done in two phases:

A. Construct the Process
B. Explore the Process

We use the name 'artefact' for the process output, which may be either product, service or system.

Step by step

A. Construct the Process

1. **Start with the end user(s)**
   Who is the main user or recipient of the artefact?
2. **Identify the main states**
   Work backwards from the artefact's point-of-use, and identify **the main states** the artefact has assumed on its journey to the point-of-use.
3. **Identify the main activities**
   Next, identify **the main activi-**  ties involved in transforming from each state to the next (in the forward flow of the process).
4. **Eliminate unnecessary states**
   Browse through each of the *states* in turn.

   i. **Is this state superfluous?**
   ii. **What conditions are neces-** sary to allow this state, or combination of states, to be removed?
   List all the outcome ideas that you discover.
5. **Reduce inefficiency**
   Browse through the *activities* of your process. How might this activity be improved, i.e. less cost, less time, better qual-  ity, ...? What conditions are necessary to facilitate an improvement?
   List all the outcomes that you discover.

PROCESS CHAIN MAPPING template   2/4

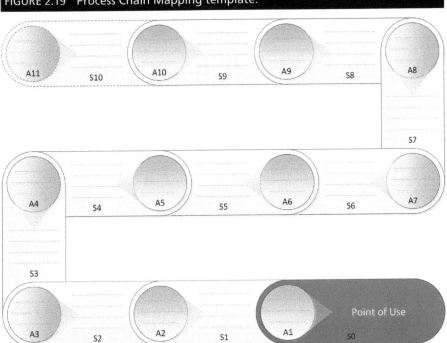

FIGURE 2.19   Process Chain Mapping template.

ACTIVITIES	
Short name (as on diagram)	**Comment**
A1	
A2	
A3	
A4	
A5	
A6	
A7	
A8	
A9	
A10	
General commentary	

STATES	
Short name (as on diagram)	**Comment**
S1	
S2	
S3	
S4	
S5	
S6	
S7	
S8	
S9	
S10	
General commentary	

FIGURE 2.20   Process Chain Mapping example – domestic laundry.

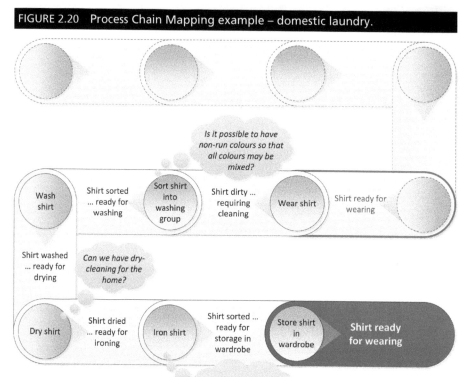

Q Why should I construct the process chain backwards?

A Constructing the process chain in a reverse flow, starting with the user, helps to keep the user as the main reference point and at the centre of attention.

Q What degree of accuracy and detail should the process chain have?

A The process chain should have between five and ten states. Fewer than five is too coarse and doesn't help your exploration of the area. Greater than ten is too detailed and will prevent you *seeing the wood for the trees*.

Q Would you say it is mandatory to use the Process Chain Mapping in all projects?

A No. Process Chain Mapping is a tool that you may use when you are looking to find innovation opportunities in an area of interest, or when you wish to prioritise target possibilities. Sometimes your target area is fully defined and you are free to focus directly on finding a solution.

Q 'Artefact' is a strange word. What does it mean?

A It means 'anything created by a human'. We use it so as to avoid having to say 'product, service or system' all the time!

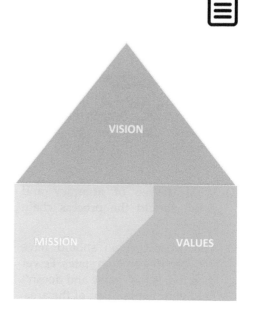

The activities of any organisation (or project) are guided and motivated by these three defining attributes.

Knowing why you are doing what you are doing (**mission**), where you are trying to get to (**vision**), and how you are going to go about it (**values**) are the glue elements that bind your organisation or project team into a coherent operating unit.

The expression of these three attributes may be intermingled, in practice. For example, values are often included in the mission statement.

Step by step

Mission
Your organisation's **mission** statement articulates its raison-d'etre, explaining why it exists, its core purpose. It is the organisation's North Star. A good mission statement informs the reader about this clearly, succinctly, precisely, and avoids bland platitudes!

Vision
Vision gives the bird's-eye view of the desired outcomes from the firm's present major strategic challenge(s). It describes a better state for the organisation, an inspiring and worthy future for all staff to strive to achieve. It is aspirational yet achievable, so that it inspires the ingenuity and coherence of effort that is necessary to build the detailed infrastructure to support the vision. The vision is often time-bound (e.g. five years), especially for internal use. A good vision statement inspires!

Values
Values describe the 'rules' by which the firm is going to accomplish its vision. The values express succinctly the behaviours and consequences of action that are desirable and those that are unacceptable.

Every organisation operates according to a set of values-in-practice, whether or not they are formally expressed. They are cross-organisational beliefs to be shared by all staff in all activities, and they guide the conduct, activities and goals of the organisation. Values are operationalised by a culture, which gives the organisation its distinctive identity.

Good values' statements guide staff how to go about their work!

MISSION | VISION | VALUES   template   2/4

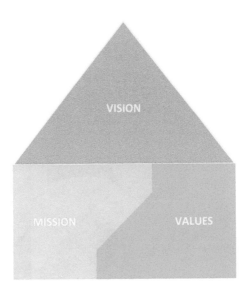

There are many different formats available, and we don't prescribe a particular one. Statements must be mutually compatible and complementary and are preferably succinct and clear.

FIGURE 2.21   Mission | Vision | Values example.

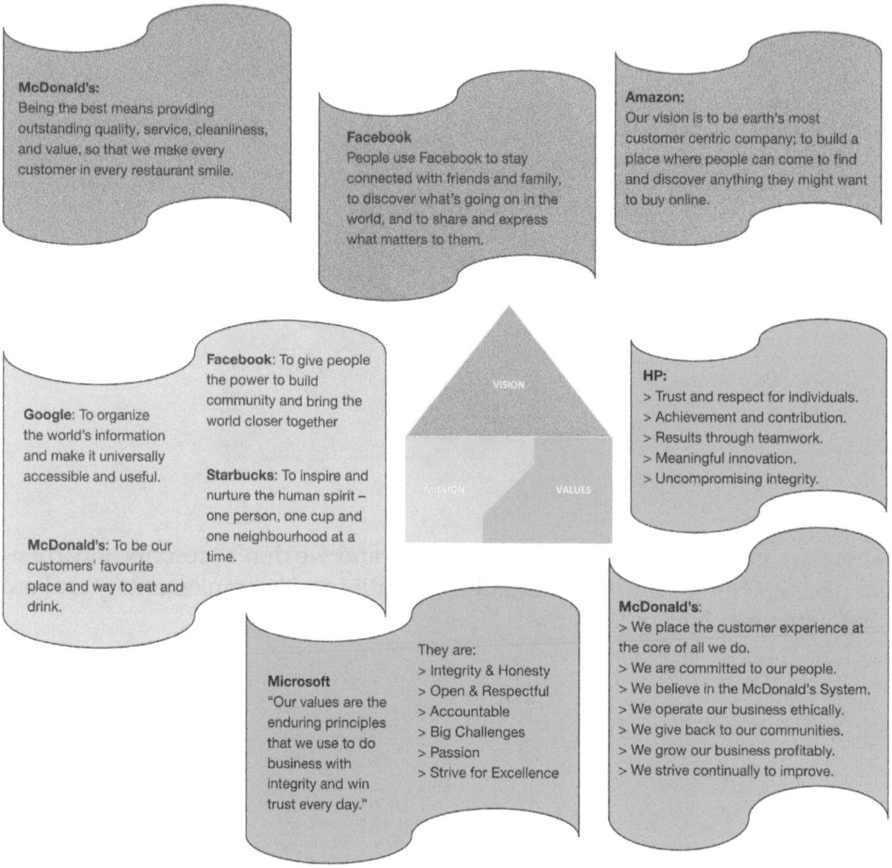

McDonald's:
Being the best means providing outstanding quality, service, cleanliness, and value, so that we make every customer in every restaurant smile.

Facebook
People use Facebook to stay connected with friends and family, to discover what's going on in the world, and to share and express what matters to them.

Amazon:
Our vision is to be earth's most customer centric company; to build a place where people can come to find and discover anything they might want to buy online.

VISION

MISSION   VALUES

Google: To organize the world's information and make it universally accessible and useful.

Facebook: To give people the power to build community and bring the world closer together

Starbucks: To inspire and nurture the human spirit – one person, one cup and one neighbourhood at a time.

McDonald's: To be our customers' favourite place and way to eat and drink.

HP:
> Trust and respect for individuals.
> Achievement and contribution.
> Results through teamwork.
> Meaningful innovation.
> Uncompromising integrity.

Microsoft
"Our values are the enduring principles that we use to do business with integrity and win trust every day."

They are:
> Integrity & Honesty
> Open & Respectful
> Accountable
> Big Challenges
> Passion
> Strive for Excellence

McDonald's:
> We place the customer experience at the core of all we do.
> We are committed to our people.
> We believe in the McDonald's System.
> We operate our business ethically.
> We give back to our communities.
> We grow our business profitably.
> We strive continually to improve.

MISSION | VISION | VALUES faqs   4/4

Q Is it necessary to have separate statements of Mission, Vision and Values?

A Many organisations don't! In many firms' statements, it is hard to distinguish between each of the three. However, we believe having separate statements is a good discipline and leads to clarity for all staff and stakeholders.

Q How many values do you recommend?

A Two or three is too few, and cannot cover the breadth of circumstances of a firm's operations. Twenty is too many, because each cannot receive sufficient mind-share and support. *You can't see the wood for the trees.* Six (plus/minus one) is a good number for such lists.

Q What style of language should I use, given that they are so important?

A These are purposeful communications, ideally for casual, familiar reference. Therefore, the language should be clear, everyday, easy to read.
Be specific, not generic.
Use phrases or short sentences; do not use single words, which are meaningless, nor long paragraphs which are hard to digest.

Q Who is responsible for defining the Mission | Vision | Values?

A This is the senior management's responsibility, from where purpose, direction and behaviour are driven. Of course, senior management must note and refer to what happens in practice throughout the firm, and refrain from handing down edicts *from on high* that are unachievable!

Before getting deep into a major innovation project it is important to understand the firm's appetite and capacity for innovation. What is the present innovation intensity within the firm and what is the ambition?

It may be that you know this well already. Even so, it is useful to refresh the key points with the team. More likely, especially with SMEs, the Strategic Innovation Profile of the firm is known only partially or not explicitly. It may be fluid and hard to pin down. Worse, there may be different understandings across different functions in the firm and at different executive levels.

We recommend that you and your team prepare a summary profile of your present understanding of the Firm's Strategic Innovation Profile. Then, make sure that you carry out a sanity check with one or more senior executives and other colleagues. The objective is to ensure strategic alignment with the firm's capabilities and ambitions at an early stage, so as to avoid disappointment or conflict later.

Some organisations have existing systems and procedures for strategic analysis. If you have access to such information, great. If you do not, we provide a simple template to get you started with a high-level overview.

Step by step

1. **Draft the template: Strategic Innovation Profile**
   Using whatever resources are available, complete the template as best you can. This is not a creative exercise, so there is no need to involve a full team. Perhaps you might bring in a colleague who, you guess, is familiar with the organisation's strategy and overall innovation activities. This is your draft profile.

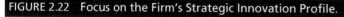

FIGURE 2.22   Focus on the Firm's Strategic Innovation Profile.

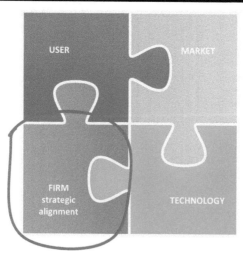

2. **Explore further and complete the profile**

   Who will help you give more authority and credibility to the profile? Perhaps the CEO will oblige, or another senior executive. Certainly, you need to get the perspectives of Marketing and Engineering functions, or equivalent.

   You may uncover differences of understanding, either nuanced or substantial. Be careful! Try to avoid the urge to be a good corporate citizen by bringing these divergences to general attention for resolution. Remember that your main responsibility is to the project and such corporate level discussions are sure to be a great way to eat up time, and perhaps lose allies. You should attempt to resolve anomalies only to the extent that clearly they have potential to have major impact on the project.

3. **Socialise the profile with the team**

   At the project kick-off meeting or soon afterwards, present the profile to the team and invite discussion. It is likely that there will be differences of opinion and even heated debate. This is okay.

   The real purpose of the exercise is to bring these important 'environmental' topics into the mind so that they can be properly considered during the innovation development. The forthcoming innovation might require substantial changes to some elements of the present profile, such as mandating new competencies or vision and brand positioning changes. It is important for your team to know what you will propose may meet resistance, and hence prepare to counter this, rather than to sleep-walk into a cul-de-sac of resistance.

FIRM'S STRATEGIC INNOVATION PROFILE template 2/4

FIGURE 2.23 Firm's Strategic Innovation Profile template.

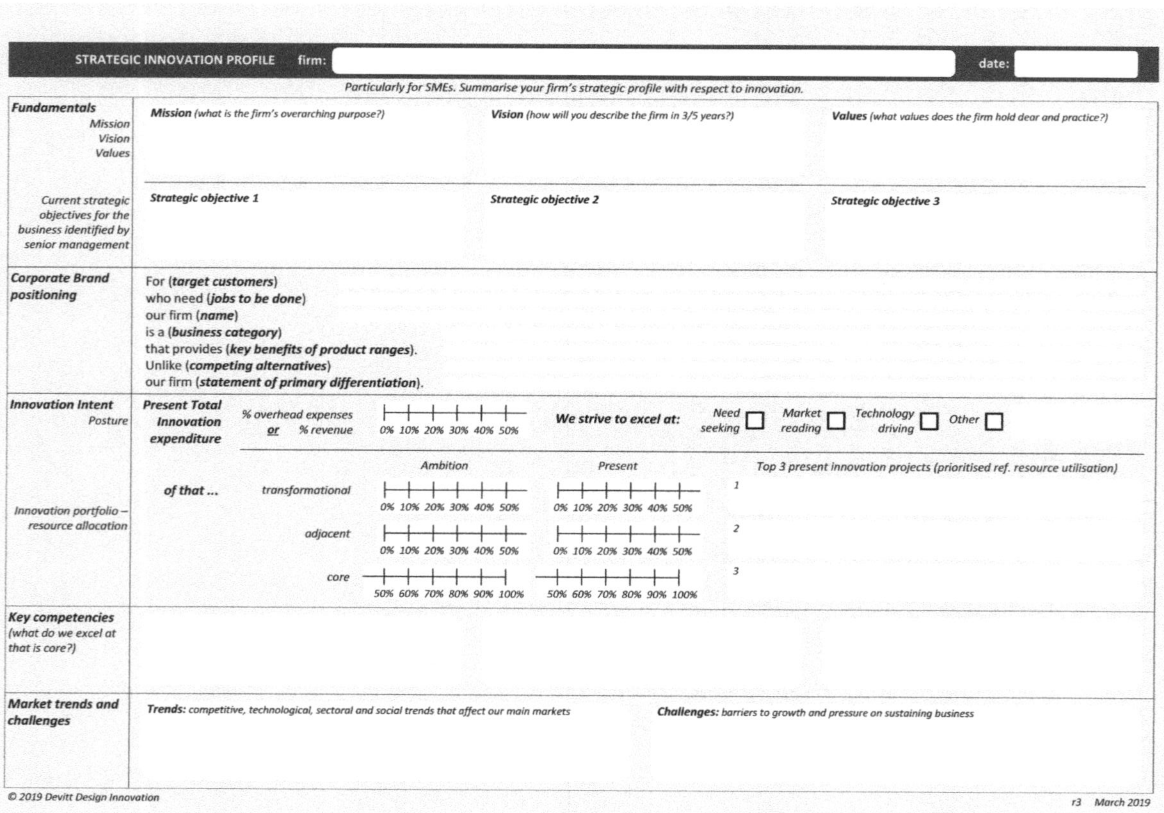
STRATEGIC INNOVATION PROFILE firm: date:

Particularly for SMEs. Summarise your firm's strategic profile with respect to innovation.

Fundamentals
Mission
Vision
Values

Mission *(what is the firm's overarching purpose?)*
Vision *(how will you describe the firm in 3/5 years?)*
Values *(what values does the firm hold dear and practice?)*

Current strategic objectives for the business identified by senior management

Strategic objective 1
Strategic objective 2
Strategic objective 3

Corporate Brand positioning

For (**target customers**)
who need (**jobs to be done**)
our firm (**name**)
is a (**business category**)
that provides (**key benefits of product ranges**).
Unlike (**competing alternatives**)
our firm (**statement of primary differentiation**).

Innovation Intent
Posture

Present Total Innovation expenditure
% overhead expenses *or* *% revenue*
0% 10% 20% 30% 40% 50%

We strive to excel at: Need seeking ☐ Market reading ☐ Technology driving ☐ Other ☐

Innovation portfolio – resource allocation

of that ...

	Ambition	Present
transformational	0% 10% 20% 30% 40% 50%	0% 10% 20% 30% 40% 50%
adjacent	0% 10% 20% 30% 40% 50%	0% 10% 20% 30% 40% 50%
core	50% 60% 70% 80% 90% 100%	50% 60% 70% 80% 90% 100%

Top 3 present innovation projects (prioritised ref. resource utilisation)
1
2
3

Key competencies
(what do we excel at that is core?)

Market trends and challenges

Trends: *competitive, technological, sectoral and social trends that affect our main markets*
Challenges: *barriers to growth and pressure on sustaining business*

r3 March 2019

FIRM'S STRATEGIC INNOVATION PROFILE example 3/4

FIGURE 2.24 Firm's Strategic Innovation Profile example.

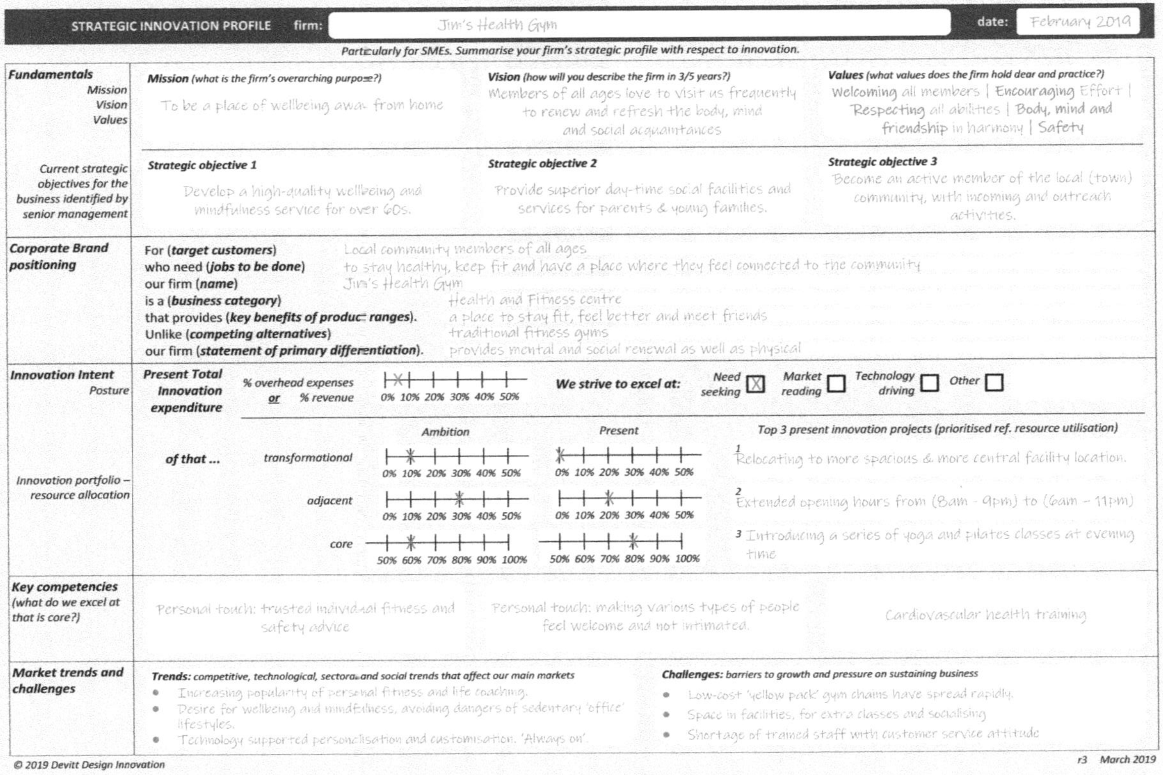

STRATEGIC INNOVATION PROFILE firm: Jim's Health Gym   date: February 2019

Particularly for SMEs. Summarise your firm's strategic profile with respect to innovation.

Fundamentals
Mission
Vision
Values

Mission *(what is the firm's overarching purpose?)*
To be a place of wellbeing awa. from home

Vision *(how will you describe the firm in 3/5 years?)*
Members of all ages love to visit us frequently to renew and refresh the body, mind and social acquaintances

Values *(what values does the firm hold dear and practice?)*
Welcoming all members | Encouraging effort | Respecting all abilities | Body, mind and friendship in harmony | Safety

Current strategic objectives for the business identified by senior management

Strategic objective 1
Develop a high-quality wellbeing and mindfulness service for over 60s.

Strategic objective 2
Provide superior day-time social facilities and services for parents & young families.

Strategic objective 3
Become an active member of the local (town) community, with incoming and outreach activities.

Corporate Brand positioning

For *(target customers)* who need *(jobs to be done)* our firm *(name)* is a *(business category)* that provides *(key benefits of product ranges)*. Unlike *(competing alternatives)* our firm *(statement of primary differentiation)*.

Local community members of all ages
to stay healthy, keep fit and have a place where they feel connected to the community
Jim's Health Gym
Health and Fitness centre
a place to stay fit, feel better and meet friends
traditional fitness gyms
provides mental and social renewal as well as physical

Innovation Intent
Posture

Present Total Innovation expenditure % overhead expenses **or** % revenue

Ambition
0% 10% 20% 30% 40% 50%
50% 60% 70% 80% 90% 100%

Present
0% 10% 20% 30% 40% 50%
50% 60% 70% 80% 90% 100%

We strive to excel at:
Need seeking ☒ Market reading ☐ Technology driving ☐ Other ☐

Innovation portfolio – resource allocation
of that ...
transformational
adjacent
core

Top 3 present innovation projects *(prioritised ref. resource utilisation)*

1 Relocating to more spacious & more central facility location.

2 Extended opening hours from (9am - 9pm) to (6am - 11pm).

3 Introducing a series of yoga and pilates classes at evening time.

Key competencies
(what do we excel at that is core?)

Personal touch: trusted individual fitness and safety advice

Personal touch: making various types of people feel welcome and not intimidated

Cardiovascular health training

Market trends and challenges

Trends: *competitive, technological, sectoral and social trends that affect our main markets*
• Increasing popularity of personal fitness and life coaching
• Desire for wellbeing and mindfulness, avoiding dangers of sedentary office lifestyles.
• Technology supported personalisation and customisation. Always on

Challenges: *barriers to growth and pressure on sustaining business*
• Low-cost 'yellow pack' gym chains have spread rapidly.
• Space in facilities for extra classes and socialising.
• Shortage of trained staff with customer service attitude

© 2019 Devitt Design Innovation   r3   March 2019

FIRM'S STRATEGIC INNOVATION PROFILE   faqs   4/4

Q You say this is for SMEs. Why only SMEs?

A Larger firms, commonly, do strategic analysis regularly, and produce data and summaries that are easily accessible.
In many SMEs, the situation is more fluid and information is not readily accessible. You can tie down this fluid strategic status, at least partially and for your own purposes, through generating the Strategic Innovation Profile.

Q Surely, these strategic matters must be the concern only of senior execs. Is it necessary for all team members to spend time on this?

A Strategic alignment of any project is vital! Everyone in the team should know the broad direction that the innovation project is travelling. It is worth a few hours for everyone to understand this alignment from the start.

Q In our firm, it's almost impossible to get accurate information. What should we do?

A Somewhat intriguingly, we find the accuracy of the end product (Profile Canvas) is not the most important part of the exercise. The strategic elements of most firms are fluid or subject to interpretation to some degree.
However, the greatest value lies in thinking deeply about all the parameters of the Strategic Profile so that team members become better informed about relevant issues, prospective difficulties and possibilities.

FIGURE 2.25   "Plans are worthless, but planning is everything." Dwight D. Eisenhower.

The innovation team's starting point for innovation is often an incomplete vision or brief. More often, the outline of intent is fuzzy, giving general direction rather than specific instructions. What does this outline intent imply? What does it entail? What is its scope? What are the big issues? Where do you start?

Every project must have direction and the Initial Project Outline provides this. However, innovation is a special learning process and it should not be overly constrained too early, particularly for breakthrough innovation efforts. Discovery of new ways to add value, which are not foreseen or foreseeable from the start, usually happens organically when the team has freedom to explore new learning opportunities. Hence we emphasise the word initial in the Initial Project Outline.

It starts off as a guide but should be considered to be malleable and amendable in response to new learning uncovered throughout the early stages of the project, in particular. The Initial Project Outline also incorporates the primary motivation for the project with possible areas of interest identified in addition to other scoping parameters.

At times, we have called this a 'scoping' document. But this term has implications of prescriptive finality, and fails to acknowledge the more malleable nature of a truly transformational innovation project. We emphasise the term 'Initial' to give emphasis to the expectation that changes will, indeed should, come as the project progresses.

Step by step

1. **Complete the template: Initial Project Outline**
   Using whatever resources are available to you, complete the template as best you can. This is not an exercise in definitive prescription, so there is no need to involve a full team, at first. Perhaps you might bring in a colleague who, you guess, is familiar with the origins of the project and of the firm's expectations from it. This is your draft profile.

   Be careful not to be too conservative. It is better at the start to be over-inclusive of a wider scope and be ready to restrict it later, if a good argument is presented. It is less tempting or easy to expand scope once the project has started.

2. **Develop the Outline with your team**
   As well as offering other perspectives on the project, your team's spirit will benefit from contributing to development of the ideas in the Initial Project Outline and feeling the sense of ownership. The team's input is especially valuable in estimating the initial vision for the project outcome and the metrics by which it will be assessed. Refine the Initial Project Outline at a one-hour team meeting.

3. Explore further and iterate

Who will help you give more authority and credibility to the Initial Project Outline? Perhaps the CEO will oblige, or another senior executive. Certainly, you need to get the perspectives of Marketing and Engineering, or other functions. You may uncover differences of understanding, either nuanced or substantial. Be careful not to over-react to one person's opinion (even the CEO, if possible!), which may be given without sufficient consideration! You will know that you have stretched the scope and vision far enough when some people express concern rather than pleasant surprise, and appear to have their credulity stretched.

Update your Initial Project Outline as a starting guide for your project. Keep updating as you learn more throughout the progress of the project.

INITIAL PROJECT OUTLINE   template   2/4

FIGURE 2.26   Initial Project Outline template.

INITIAL PROJECT OUTLINE firm:	date:

Document your initial thoughts below. Use your best estimate as far as possible. Don't hold back! This is not a rigorous specification document. It is a documentation of your initial thoughts, estimates and assumptions, which will be evaluated and probably (hopefully) changed in the course of your project work.

Project Description	Briefly describe the circumstances that give rise to the project that you are working on. What are the main issues or dilemmas to be resolved? From where does the project get its impetus and legitimacy? (e.g. CEO, Marketing, strategic plan, technology threats, …) Background What are the high level or strategic goals (list three in order of priority) What problem(s) are you trying to solve? (list three in order of priority) Give your project a name:
Likely users and stakeholders (of your problem's solution or a new innovation)	End users? Buyers? Channel partners? Government or public agencies? Colleagues? Are they likely to be affected directly or indirectly?
Vision (desirable outcome)	Give the traits of a desirable outcome, as you imagine them at this stage. e.g. the solution must be … aesthetically appealing; compact; easy to use; have 12-month payback, set a new business direction, etc. Give traits – not features!
Success measure (how might you evaluate success?)	

r4   Sept 2019

INITIAL PROJECT OUTLINE example 3/4

FIGURE 2.27 Initial Project Outline example.

INITIAL PROJECT OUTLINE	firm:	ExamSupport	date:	Sept 2019

Document your initial thoughts below. Use your best estimate as far as possible. Don't hold back! This is not a rigorous specification document. it is a documentation of your initial thoughts, estimates and assumptions, which will be evaluated and probably (hopefully) changed in the course of your project work.

Project Description	Briefly describe the circumstances that give rise to the project that you are working on. What are the main issues or dilemmas to be resolved? From where does the project get its impetus and legitimacy? (e.g. CEO, Marketing, strategic plan, technology threats, ...)

Background

ExamSupport is a small online educational company providing online learning and tutorials for students preparing for national school-leaving examinations, with over 10 years in business, brand recognition is high. However, the old business model is now showing strategic cracks that require urgent attention. ExamSupport's business model has evolved towards selling in bulk to Telcos, who in turn use the service as new-business and anti-churn incentives (free) for its broadband customers. Hence, direct online sales of ExamSupport dwindled. Now, ExamSupport is precariously over-dependent on Telcos.

What are the high level or strategic goals (list three in order of priority)

- *find a new market(s) that is adjacent (functionally or geographically) to the existing market that is dominated by the Telcos.*
- *Use our existing capabilities and resources as much as possible, yet be commercially distinct.*
- *Time scale is urgent – less than 12 months to first deployment.*

What problem(s) are you trying to solve? (list three in order of priority)

- *How might we compete against (or utilise) the great volume of generic content on the web, which is close to our users' interest area, in order to deliver a more targeted, custom service to our user?*
- *How might we upgrade our content delivery mechanism to align fully with mobile/multi-platform use?*
- *How might we upgrade our in-house technology to accommodate greater content production efficiency?*

Give your project a name: DirectEd

Likely users and stakeholders (of your problem's solution or a new innovation)	End users? Buyers? Channel partners? Government or public agencies? Colleagues? Are they likely to be affected directly or indirectly?

End users: open – could be school leavers in other territories; or, professional certification students, or other ...

Channels: open – to be explored; could be direct online, other Telcos, other ...

Vision (desirable outcome)	Give the traits of a desirable outcome, as you imagine them at this stage. e.g. the solution must be ... aesthetically appealing; compact; easy to use; have 12-month payback, set a new business direction, etc. Give traits – not features!

- *Modern, interactive technology, building on existing content, all as a cloud service.*
- *Easily scalable and adaptable (website skinning and content production) for different markets and private labelling.*
- *Compatible with (but not necessarily central focus on) social interaction.*
- *Operation is embedded into users' lives, not sporadic, e.g. once-per-day, engagement*

Success measure (how might you evaluate success?)	• *Content repeat usage (at least 1 repeat visit per week): >100 in first month; >20,000 in 6 months* • *User sign-up rate (tbd: >nx1000 in first months; >nx100,000 in 6 months)* • *Social virality – referral rate tbd*

r4   Sept 2019

INITIAL PROJECT OUTLINE faqs   4/4

Q The Initial Project Outline seems loose compared to many project briefs. Why is this?

A The tightness of specifications at a project's early stage determines the scope for innovation throughout the project. Innovation is a learning process. At one extreme, if you are very tightly constrained from the beginning, it leaves no room for exploiting new learning that you gain as the project progresses. The specification has been fixed pre-learning.

On the other hand, of course, if the specification is too loose – 'just innovate' – then this is almost impossible. Creativity needs some constraints and direction. However, these may be 'loosely' constructed.

We favour tending towards the latter condition, where possible. Here, the early scope is left as open and high level as possible, thus encouraging possibilities for transformational innovation to come to the surface from breakthrough research insights as they arise.

In many circumstances, such openness may not be possible, and constraints are necessary. If so, then our rule is simple: keep constraints as high level as possible at the start.

Q In the Vision, it mentions 'traits, not features'. What's the difference?

A By 'trait' we mean a distinguishing quality or characteristic, described at a high level. By 'feature' we mean a particular instance, component or function described at a more detailed level.

For example, a new house design project might specify a trait of 'easy access and mobility' or 'easy opening, hands-free doors'. A particular instance of this might be the feature 'electrically powered doors activated by proximity sensors'. Or, an alternative feature might be 'doors with low friction hinges and kick-panels'.

Q Have you any advice on establishing success measures?

A Firstly, they are an indispensable guide to your progress and you need them. You might think of input measures and output measures. You must address both.

Input measures query if your evolving solution satisfies the Vision traits as specified. Output measures query if your final solution solves the problem or does the job it is supposed to do at a satisfactory level.

BIBLIOGRAPHY

Cetindamar, Dilek; Phaal, Rob; and Probert, David (2010). *Technology Management: Activities and Tools*. Palgrave Macmillan.

Christensen, Clayton (1997). *The Innovator's Dilemma*. Harvard Business Review Press.

Cohen, Wesley and Levinthal, Daniel (1990). Absorptive Capacity: A New Perspective on Learning and Innovation. *Administrative Science Quarterly*.

Cooper, Robert (2017). *Winning at New Products*. Basic Books.

Duan, Lily; Sheeren, Emily; and Weiss, Leigh (2014). *Tapping the Power of Hidden Influencers*.   https://www.mckinsey.com/business-functions/organization/ our-insights/tapping-the-power-of-hidden-influencers.

Gadiesh, Orit and Gilbert, James L. (1998a). How to Map Your Industry's Profit Pool. *Harvard Business Review*.

Gadiesh, Orit and Gilbert, James L. (1998b). Profit Pools: A Fresh Look at Strategy. *Harvard Business Review*.

Kim, W. Chan and Mauborgne, Renée A. (2005). Blue Ocean Strategy – from Theory to Practice. *California Management Review*. Spring, Vol. 47, No. 3.

Kim, W. Chan and Mauborgne, Renée A. (2014). *Blue Ocean Strategy*. Harvard Business Review Press.

Kim, W. Chan and Mauborgne, Renée A. (2017). *Blue Ocean Shift – Beyond Competing*. Macmillan.

Knapp, Jake (2015). *Sprint – How to Solve Big Problems and Test New Ideas in Just Five Days*. Bantam Press.

Liedtka, Jeanne (2015). Perspective: Linking Design Thinking with Innovation Outcomes through Cognitive Bias Reduction. *Journal of Product Innovation Management*, Vol. 32, No. 6, pp. 925–938.

Nagji, Bansi and Tuff, Geoff (2012). Managing your Innovation Portfolio, *Harvard Business Review*, May.

Porter, Michael (2008). The Five Competitive Forces That Shape Strategy. *Harvard Business Review*, January.

RAND Corporation (2020). *Delphi Method*, viewed 19 May 2020. https://www.rand. org/topics/delphi-method.html.

Schilling, Melissa A. (2010). *Strategic Management of Technological Innovation*. McGraw-Hill.

Verganti, Roberto (2016). *Overcrowded*. MIT Press.

Research

Without Research you get me-too products because you have used me-too data.

The Audit phase has provided a broad understanding of the environment for the proposed innovation, both inside and outside the firm.

Now, in the Research phase, you apply a microscope to study the people involved in the target context. Our Research focuses on acquiring a rich understanding of the needs, jobs, behaviours and aspirations of users and other stakeholders. Deeply human-centred insights are the Holy Grail because there is a good chance you may leverage them to create a great, distinctive and uniquely compelling innovation for your target.

There is a wide range of ethnographic, jobs-to-be-done (jtbd) and other qualitative methods that you may employ. The outcome is a distinctive understanding and empathy for users' circumstances, their experiences and aspirations.

FIGURE 3.1 Chapter 3 outline.

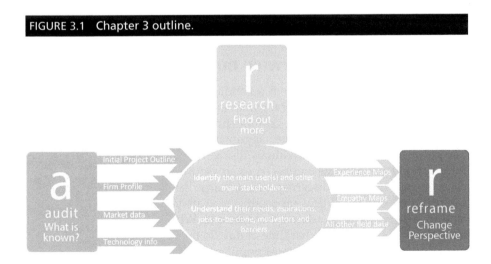

FIGURE 3.2   The scope of Research.

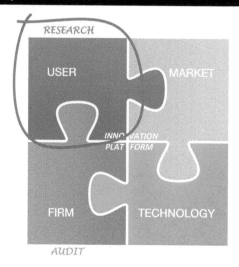

PEOPLE HAVE PROBLEMS

In the foundational Audit phase, you used the firm–market–technology frame-work as a guide to explore your innovation focus area and its environmental characteristics. This Research chapter and the Research phase of your innovation project are dedicated to understanding the *user* dimension, an asset that will make your product uniquely compelling for the targeted user.

When you put the people who will use and interact with your innovation at the front and centre of your investigations, it is an acknowledgement of their importance. People, whose actions and interactions are governed by psychological and social complexity, are the most challenging and difficult-to-understand element of every innovation system. Ultimately and always, only people have problems or aspirations. Machines do not! Machines have features that may be the cause of problems or their solution. But, to have a problem requires consciousness with an unfulfilled aspiration. Additionally, it is the user or some other interested person who will be the arbiter of the innovation project's success or failure.

'WHAT' QUESTIONS AND 'WHY' QUESTIONS

You may learn about people from secondary research, but this is inevitably limited and incomplete. When you gather information at your desk computer, from published reports or even custom surveys, you often get a view of the subjects that is distant, anodyne, aggregated and dehumanised. From the perspective of your desk, individuals lose personality and become statistics. Information loses richness and becomes an assembly of bare data points with consequent difficulty to discern patterns, if any. With enough data points and sophisticated data-analytical processing, you may glean some insights that tend to answer 'what' questions. For example: What is the preferred refresh-ment drink of teenagers? What time of day do they tend to buy most refresh-ment drinks? What outlets are most popular for casual, single quantity drinks purchases, and where are the bulk purchases made?

It should go without saying that knowing facts is fundamental. It is necessary and valuable. However, alone, quantitative data rarely inspire.

Deeper understanding and problem-solving inspiration come from knowing the causes behind the facts, i.e. the answers to the 'why' questions. Knowing the crystallised facts is the start; it is a state where you know as much as, but not much more than, a competitor who might access the same industry reports and Google searches. Occasionally, you might uncover from a quantitative survey a particularly interesting pattern or insight but, usually, it is just not enough to give you clear direction or a distinctive advantage. In order to succeed at transformative innovation, you need to get a richer and more nuanced understanding of the users' behaviours, aspirations, needs and desired outcomes.

LET THE DISCOVERY BEGIN!

Like the old-time pioneering gold miners, after the preparation work of Audit, now you are suited, booted, tooled-up and ready to explore for gold. So, let's get to it! This is where the fun starts. It is where you meet people you have never met before and see things through others' eyes – like you have never seen them before. Here, you will discover insights that change your own perspective such that your world will never be the same again!

Does that sound overly dramatic? Yes, it might be an exaggeration for some cases. However, often this truly reflects the excitement and fulfilment that you get from mining nuggets of knowledge that might lead to transformative innovation. These golden insights, once discovered, cannot be unseen. Once understood, they cannot be forgotten, and they will guide you to inspiring innovation possibilities.

ENABLING DIFFERENTIATION

Understanding the people that own the problem is the key enabler of differentiation. You will never regret time spent getting to know what people are really trying to achieve, what frustrates them, inspires and motivates their behaviour. Through this effort, you acquire a proprietary perspective with a distinctive potential to make your new product different, outstanding and especially compelling to the user, in a way that no one before you ever envisaged. Without user research, you will have me-too new products because you have used me-too data. The only differentiation possible in the marketplace would be in price unless it is fuelled by proprietary, differentiated factors or processes of production, which are difficult to achieve. Yet, price competition is a scary road to travel.

The objective of researching people is straightforward. You are seeking to achieve a complete understanding of the behaviours, motivations and emotions that surround 'the context of interest'. This is encompassed by the term *empathy*, which is a term adopted by design innovation practitioners to describe a prime objective of their human-centred research.

The imperative for achieving empathy research of high quality is to "get out of the building". You cannot avoid it, even though it may appear scary at first. "Get out of the building" is the clarion call phrase of Steve Blank, originator of the Customer Development process and father of the Lean Startup process, which we will talk more about later in this book. Within the building, your desk research of market, technology and trends during Audit investigations has exhausted all inspiration and insight to be gained there.

INSIGHTS

We refer regularly to the search for insights. By *insight*, we mean identification and a deeper understanding of causal connections, usually uncovered or inferred from immersion and empathy. In the words of renowned psychologist in the field of intuition and insight, Gary Klein, an insight is an unexpected or unpredicted shift to a better story. Klein identifies three paths to insights with respective triggers of *contradictions*, *connections* and *creative desperation* (Klein, 2013). The first two of these are especially relevant to the *Research* activity.

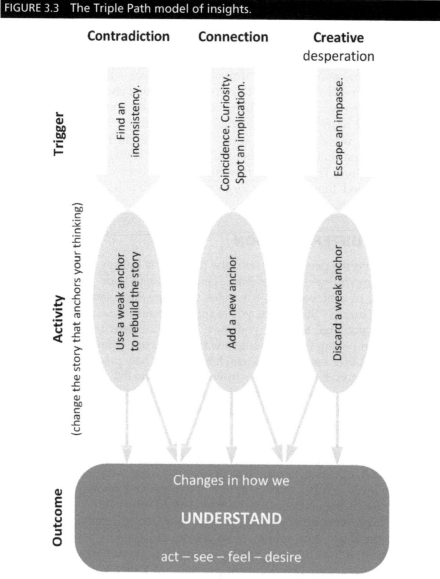

FIGURE 3.3   The Triple Path model of insights.

Source: Klein, 2013.

Special insight requires special intimacy, which you acquire only through ethnographic field research work, a methodology derived from the field of Anthropology. Of course, the principle is a well-established rubric in many cultures. An old Irish proverb says: "You won't know people until you live with them" (*Ní aitheantas go haontíos*).

THE HUMAN-CENTRED INNOVATION QUESTION

We rely heavily on ethnographic methods drawn from the discipline of Anthropology. In its pure form, Anthropology is a social science that seeks to understand people in the various contexts of their lives. However, we should not forget that, as innovators, our ultimate aim goes further. We are seeking to build on the understanding so as to create an innovative intervention that will add value to their lives, which they will feel compelled to adopt.

Why will a person buy our new product? How will we encourage target prospects to become real customers? The eternal marketing questions are the constant focus of market researchers, product managers and marketing communications teams. At this Research stage, we ask a different question, which is upstream yet correlative:

What prospective change in a person's circumstances would be so attractive as to compel them **to acquire the means likely** to bring about that change?

Note that we are not looking for the means (solution) at this stage. We are looking for areas of imperfection in people's lives – from their perspective – that we conclude they would like to change if they had the means to do so. There are many terms used by different authors to describe this basic approach, e.g. design innovation and user-centred innovation both seek to determine and prioritise user needs, jobs-to-be-done, and similar terms.

There may be many 'solutions' to give effect to the innovation, and some will be more compelling and effective than others. Innovation is not confined to a new product that performs a mechanical function. It may also be a service or a system that performs an emotional or social function equally. It may have a combination of multiple of these. In fact, the best innovations are compellingly attractive to the user at multiple levels. They perform multiple functions – the primary one and subsidiary ones – across the physical, emotional and social domains. As an example, much of the success of Airbnb, the online accommodation sharing business, is due to offering lower cost accommodation to travellers, combined with the opportunity to meet interesting people and see quirky properties, in a way that allows buyers and sellers to establish mutual trust, all of which is a much more attractive package to many than a bland vanilla hotel experience.

It is surprising how some innovators are so hesitant to engage deeply in human-centred research. For many, especially those who have engineering, IT, scientific or business management backgrounds and have been immersed in quantitative methods up to now, understanding human behaviour is a foreign, fuzzy and forbidding territory. It is a journey of exploration where the outcome is uncertain because there is not a single, unique target to be aimed at. The outcome of human-centred research evolves, morphs and crystallises in the mind of the researcher as the research proceeds. Conversely, however, it is precisely in this journey of discovery and evolution that the joy of human-centred research also lies. The initial uncertainty masks an array of awaiting discoveries.

BEING A RESEARCHER

Humility is a particularly important attribute in carrying out qualitative human-centred research, and it is not for reasons of altruism or morality. When a researcher acknowledges from the outset her limited understanding, she frees herself to be more open to discovering, recognising and welcoming subtle clues in the research that may lead to the much-sought-after golden nuggets of insight. A strong, firmly held opinion (or bias) can be disabling when it is (inevitably) selectively entertained, prioritised and bolstered at every step.

As well as humility, it is important for the researcher to be aware of being in 'measurement stealth mode'. The atomic physicists among our readers will know of Heisenberg's Uncertainty Principle, which asserts as a fact of nature that the act of measurement *always* introduces a disturbance to the parameter being measured. The fundamental truth is that you cannot, for example, measure the temperature of an oven without disturbing the temperature from what it would have been without making the measurement. Business and psychology students will know this as the Demand Effect (also sometimes considered as one interpretation of the more famous Hawthorne effect), where, when a subject under study figures out that she is actually being measured, she changes her behaviour to align with her interpretation of the expected measurement outcomes.

Hence, the skill in making good measurement is to minimise the disturbance in order to get as close as possible to 'true' data. This is equally important in 'measuring' humans. The acts of observing, interviewing, surveying and all other research methods inevitably interfere with how the subject would have behaved or thought if the research had not been done. Once the researcher has moved on, the subject is likely to revert to the pre-research status quo. If you want to get the truest and most accurate information from your research efforts, you must constantly take care to be as non-interfering as possible, while still getting the information required. The subtlety of appropriately balancing non-disturbance against the need to access the right data and acquire sufficient data is a skill that requires ingenuity and improves with practice and experience.

If you are new to this type of research, we encourage you not to be reluctant: follow the Nike slogan and *just do it*. But, start simply with some practice runs in a non-critical setting. Maybe, spend some time simply observing people in the local shopping mall. How do they behave? What catches their attention? When are they confused? How and when do they interact? Talk with them casually about their experiences. Practise listening, and speak only for the purpose of drawing out more comments from the person you are addressing. Practise your notetaking and reflect on your conversations to try to get a deeper understanding of what they may be thinking or what may be at play. Speak with others to see if you can get verification of your initial insights.

There are many guides to help you in your observations. One of the seminal and most famous is Spradley's nine dimensions of descriptive observation (Spradley, 1980). Dr James Spradley published the framework to assist

Table 3.1 James Spradley's nine dimensions of participant observation.	
Space	layout of the physical setting; rooms, outdoor spaces, etc.
Actors	names and relevant details of the people involved
Activities	the various activities of the actors
Objects	physical elements: furniture, etc.
Acts	specific individual actions
Events	particular occasions, e.g. meetings
Time	the sequence of events
Goals	what actors are attempting to accomplish
Feelings	emotions in particular contexts

ethnographic researchers by lightening the burden of cognitive load associated with simultaneous data collection, analysis and narrative construction.

The nine dimensions are shown in Table 3.1.

EXPLORING BEYOND THE OBVIOUS

So, how does this work in practice? Say you are interested to learn how a busy office worker, Jane, experiences a bagel bar around mid-morning in a busy office district. Obviously the central tangible experience is when Jane enters the bar or joins the queue for service. The bagel is served and eaten in-house or taken for consumption elsewhere. This is the core activity and your job is to get a detailed understanding of everything surrounding it. You must zoom into particular details such as menu clarity, speed of service, presentation of the food and hundreds of other details. But, there's a lot more. You also must zoom out and understand why Jane chose this Bagel Bar to buy in, or why Jill (Jane's office colleague) did not choose this bar but instead chose another deli down the road. These are the lead-in or antecedent conditions for their mid-morning food encounters. Subsequently, Jane eats her food. Does she consume other food or drink with it or has she bought the full repast from the Bagel Bar? Does she enjoy it? Is she satisfied?

But, pause for a while, and zoom out even further. Why are Jane and Jill doing this, really? You might first answer: they are buying food. More broadly, you might say they are satisfying hunger. Alternatively, maybe their main goal is to take a break from the office for 15 minutes. Or, perhaps their real interest is to meet with other regulars for a social chat and keep up to date with the latest gossip. Possibly, you could describe their motivation as a multi-layering of all of these. We often make use of the 5 Whys technique, which we describe in more detail later in the methods section of this chapter. See also the side bar summary of the 5 Whys technique.

Exploring beyond the immediately obvious is the researcher's key skill and this is where valuable insights may be discovered. As well as exploring before and after the core job, the researcher must alternately zoom in and zoom out to get a full understanding of what the user-under-study is trying to achieve, both at an overview level and at a detailed level, even though this aspiration often may be a subconscious one. See Figure 3.4, which represents the multiple modes of research that are necessary.

Besides your target user's aspirations, you also want to find out what are her key motivators to action and barriers to fulfilling those aspirations.

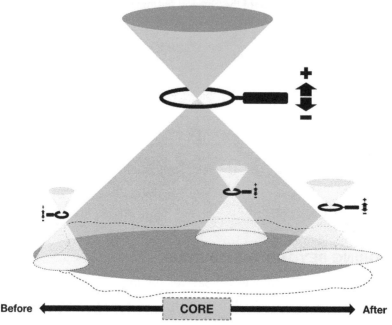

FIGURE 3.4   Scope of user research: zoom in and zoom out.

In addition, so as to guide you to an impactful and commercially successful innovation, you especially need to understand the user's particular desired outcomes from the innovation. How will they be measured and what metric level defines success?

5 Whys

In the mid-20th century, Toyota Car Company developed the 5 Whys method of seeking out the root cause of a problem. Since then it has become a popular component of problem solving for many practitioners.

It is best done by a team, with a clear problem statement. Asking *why* the problem has occurred elicits a suggestion from the team. This in turn is interrogated (*why?*) to distinguish if it may be a symptom of a deeper cause or is itself the fundamental, systemic or root cause.

The process repeats until the team agrees on the root cause, which usually takes three or more rounds. Then, corrective actions may be taken to avoid this root cause and promise a more favourable outcome for future situations similar to where the initial problem was experienced.

It is a method that is simple in concept and yet powerfully effective.

WHO IS YOUR USER? PERSONAS

How do you identify who it is that you should study so deeply? Good research takes time, so you have to target wisely. At the same time, you cannot be too restrictive or you may miss out on opportunity. As you homed in on your innovation area in the Audit phase, you identified probable users and other stakeholders. These are people who are active in the area, or people who are dependent on or influenced by activities of the area.

The market may exist already, with the present stakeholders easily identifiable, or you may have guessed at the prospective stakeholders of a proposed new market. In some circumstances your task may be to innovate for a defined set of users and, in this case, clearly, there is no difficulty identifying them.

Where a new market *and* new offering are planned, a chicken-and-egg situation arises. The product/offering (when it is conceived) will define the user. But, the user (when she is found) defines the product offering! In such a case, the innovation team must start by exploring the area thoroughly to discover and refine who might be target users for deeper research. Initial guesses will provide the prioritised targets for research, and iteratively this will provide increasing amounts of real data to guide convergence to a target where impact and compelling value may be added.

Throughout the research, your project team will talk with and get to know many stakeholders and gather much data, before you will be in a position to describe accurately the key characteristics of the main user(s). While each person may be a distinct individual, traditionally firms try to identify clusters or segments of the market, which are close enough in some respects to allow treatment as a homogeneous grouping, for the purpose of targeting product offerings and communications. The segmentation is often based on demographics such as age, gender, socio-economic class, education level, etc. or product feature set and performance, such as minimum features versus all bells and whistles, or amateur use versus professional use.

More recently, a jobs-to-be-done approach advocates segmentation that is based (as the name suggests) on the job that a person or persons are trying to get done. These may be principal jobs and ancillary jobs and be functional or emotional or both. In practice, the jtbd segmentation may have a loose or even no connection to traditional demographics. So, for example, an older woman and a teenage boy, in the context of being classmates at university with impending end of term examinations, may be united in trying to find focused extramural tuition, because of the teaching inadequacies of their course professor.

Design innovation research embraces the jtbd approach and goes further by giving central emphasis to a finely honed user empathy, which is maintained throughout the project. This empathy serves to maximise inspiration at ideation, and adoption of users' feedback at concept validation, in particular. The concept of a persona is an important tool that design innovators use to 'put flesh on' user empathy for the purposes of visualising, sharing and recording the data. However, it should be remembered that design innovators use personas for reasons that are different and in a fashion that is distinct from many other disciplines' use of personas.

A POWERFUL SECRET

Expert designers have a powerful secret that seems to have gone mostly unrecognised in the broader innovation community. It is, when you design to delight one person, usually the result satisfies and is compelling to many others as well. Designers use the concept of a *persona*, an artificial entity, to provide a focus for efforts throughout the project and an inspiration for solutions that will ultimately delight the user. It sounds simple, but it goes against the instinct of those trained in quantitative methods and demographic profiling. After all, they would say, the one persona targeted may be an exception! This might be a problem if it were the full story, but it is not. The persona provides focused inspiration; but it is the subsequent validation testing and updating that provides confirmation.

Further, the persona used by designers is synthesised from solid research. It is a rich construction of personal and social characteristics, aspirations, objectives, behaviours and circumstances of a fictional yet representative user. Often multiple distinct personas (up to five or six) will be used to represent the variety of users and stakeholders expected. This *persona* is firmly grounded in your research findings, and it functions as a vehicle to carry these findings throughout the project.

The close connection to research makes the persona highly effective and extremely valuable. For the reasons given above, we advocate developing personas straight after the research phase has completed. When the data is studied, analysed and filtered, it brings to the surface defining characteristics of one or multiple representative, prospective users. We describe this method in the next chapter, Reframe.

RESEARCH EXAMPLES

Let us look at some alternative approaches to researching a food and snack service business in Dublin's refurbished docklands.

What research might have led to the *Docklands Bagel Bar* coming into existence? Here are four alternative stories.

1 A well-known product with a clearly-identified and familiar user

National Bagel Retailing Company had a chain of 42 bagel shops across three major UK and Irish cities, and was constantly on the lookout for opportunities to expand. Mostly it served densely populated office areas and the recent Dublin docklands office developments fitted nicely with the profile of existing users in other locations. A well-profiled customer and proven matching products need only minimum further user research, for confirmatory purposes.

2 A new product for known users

Docklands Salad and Yogurt Company (DSYC) has been serving healthy salads for nearly four years to docklands office workers. But, over the last year, especially, business volume has stagnated and even decreased in the face of new competitor food shops. DSYC initiated a project to see how it might re-capture some of its lost clients and target substantial growth.

The project team researched existing regular users and non-users. They were particularly careful to look for people with 'extreme' characteristics in their patterns of engagement with the shop.

- They spoke with people who came every day.
- They spoke with people who came just one day per week, consistently.
- They spoke with people who bought large amounts, consistently.
- They spoke with people who bought very small amounts, consistently.
- They looked outside the shop and noticed people who consistently passed the shop and went to another food shop. They stopped these people and tried to understand why they did not come in.
- They sought out people who had once been regular customers and were no more, and they tried to understand why this was.

The observations and questioning that the project team did with these research subjects was centred around understanding what were their functional, social and emotional needs to be satisfied and outcomes they were trying to achieve.

They discovered that most of the research subjects liked to feel they were eating healthily, but many did not feel satisfied by salads. Salads did not have enough substance to maintain a feeling of fullness beyond an hour or so and the tastes were limited. Some forced themselves to put up with the residual hunger pangs once a week, usually on Mondays after the weekend splurge, which made them feel virtuous. Some tried for a while as part of a weight saving diet, but gave up after a month or two. Others really were not at all interested in salads but came simply to meet other regulars and share the latest gossip.

As a solution to address the findings of their research, reached after much deliberation and ideation, the project team proposed bagels as a food low in fat, salt and cholesterol, which is nevertheless rich in carbohydrates to keep the customer satisfied. It also lends itself to being eaten hot or cold and with a wide range of tasty fillings.

The *Docklands Bagel Bar* was born.

3 Existing product for new users

Sam's Student Lunchbagel Company was a small bagel franchise serving a large city-centre university campus with bagels and coffee. Students' priorities were affordability and simple, filling food. Despite students' reputation, Sam knew that consistently good quality was also necessary.

Over the years the adjacent docklands landscape had dramatically altered with the addition of large new office blocks and retail shopping malls, but strangely this had little influence on SSLC's business. Perhaps this was a sign of the clear focus that SSLC had in serving its traditional customer base.

But, times moved on and Sam Senior retired. Samantha Junior took over management and was looking to expand the business and perhaps diversify away from the strong identification with and reliance on student clientele, with its inevitable price-consciousness and seasonality. SSLC had an established expertise, reputation and supply chain in bagels and Samantha Junior thought she should leverage this as part of the new business development. Additionally, Sam Jr had been aware of vacant premises one and a half kilometres down the road near the heart of the docklands that had two and a half times the space of her existing shop. Sam Jr knew her task was to find new customers and to hone a new bagel offering around their requirements.

Sam Jr hired an innovation consultant to work with her company for two months to help research a solid proposal for developing the business. The consultant set about investigating the possibilities by profiling and segmenting the demographics of the people who were busy in the area around the shop every day. She also observed the eating establishments in the area and monitored which type of clients most commonly frequented them. In total, she spent three weeks conducting ethnographic research around the eating establishments frequented by the office workers. She observed the eating preferences of different groups, their lengths of stay for eat-in and whether they ate alone or in groups. She noted who was more likely to buy food for take-out and she followed these clients to see where they ate their food. Some brought it back to the office; but, when the weather was good, a great number brought it to the park to eat on a bench, by the lake or on the grass, while chatting with friends or watching the crowds.

She observed how awkward and undignified it was to eat large rolls or sandwiches while sitting on grass or on the lake wall. She talked with many of the people she observed to see why they had bought the particular choice and whether they varied their choice from day to day.

Armed with all this data, Sam Jr and her innovation consultant were able to identify some prospective target users who could be well served by the business's bagel capabilities and competences. The consultant drew up multiple rich draft persona descriptions of the key user types and the whole team spent a half-day discussing and refining them until they settled on four personas to guide their further work. Together they identified themes for exploration such as speed of serving, convenience of eating, variety of taste and health-consciousness, and they set out to develop some prototype concepts, which they would test with their target users.

The *Docklands Bagel Bar* was on the way!

4 A new product for new, unidentified users

Downtown dockland was re-invented and re-developed from derelict dockyards into modern office blocks and retail stores over the past ten years. Office construction work was intense for most of that time, and food shops proliferated to fill the tummies of those construction workers. As construction activity decreased, so did the number and the aggregate appetites of the remaining construction workers.

Pete's Pies and Pastries made good business for five years serving high-energy, satisfying foods to mostly construction workers. Pete could see the end of the line approaching within a year or two, with the existing market getting more and more competitive and reducing margins, so he commissioned an innovation research team to advise him on how he might survive in the food business. Pete considered his main assets were his knowledge of food preparation processes and standards, and his store location with a long-term lease.

The research team started by meeting Pete in his store one Tuesday afternoon, and they invited Pete to join them on a two-hour walk around the streets and malls within a 1 kilometre or so distance from his shop. During the walk they took notes, photos and had some discussions around what they observed.

Upon return, they spent another two hours mapping out in mind map fashion the observations they had made together with inferences and estimates of significant prospective user segments and their behaviours and attributes (see Figure 3.5). These were initial estimates based on preliminary observations and estimates from experience. But they provided a useful platform for further research.

The research team concluded that the area had:

- A large office population, with a growth trend because of a clear pipeline of new and soon to be completed office space.
- A large number of retail shoppers, especially between the hours of 10:30am to 6:30pm.
- A dwindling construction workforce.
- An increasing number of startup, specialist 'high-end' food suppliers.

As a next step, the team undertook to study these populations in more detail with these charging questions:

- What are the sizes of the various populations?
- What are their daily routines, if any?
- What are their preferences for food consumption?
- What are the main experiences and emotional drivers in their daily lives? How does food consumption fit into it?
- What other concerns or aspirations do they have around food or food shops, besides nutrition?

The innovation team formed three separate groups of two people and each was allocated to research one of the main demographic segments identified

FIGURE 3.5   Observation research mind map for Pete's.

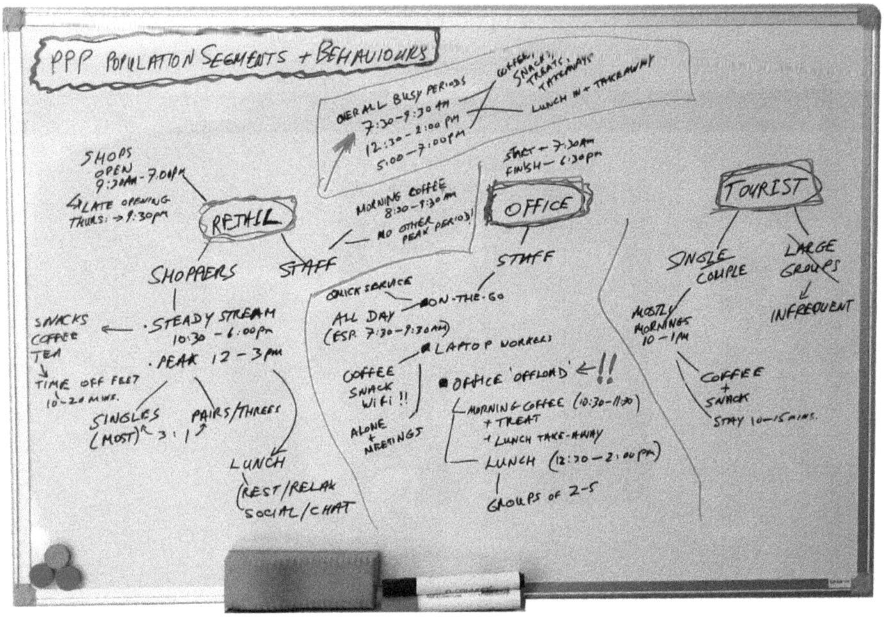

above. While each pair had a target, they also had a remit to keep their eyes wide open, especially for anything that might be exceptional (e.g. coincidental or contradictory) or did not fit neatly into the preliminary segmentation above.

They spent two weeks doing this, and reconvened with Pete the following week to report on findings:

- The office workers were the most promising segment in terms of population size and probable available expenditure.
- Retail store staff had similar requirements but probably would eat and snack at different and more spread-out times. This could be an advantage for balancing of the shop's customer load profile.
- While office staff and retail staff may appear as different segments, in fact their needs and jobs-to-be-done may have large overlaps and might be considered together.

The innovation team presented their research with the aid of three rich persona boards developed from their research data.

The innovation team and Pete agreed to proceed further with detailed ethnographic research of the office and retail shop staff in order to determine what new product proposition would provide the most compelling offering.

The *Docklands Bagel Bar* proposition was being formed.

User research – summary

The distinctive advantage that a design thinking approach gives to an innovation project is that it extends the platform for innovation, beyond the more traditional considerations of market, technology and firm, to include an empathic understanding of the people involved. This leads directly to more powerful human-centred user insights that can leverage transformative innovations.

Who is your targeted main user? Who are the central and other stakeholders that you must know so as to develop a great product?

How can you know them better? What are the dominant circumstances, needs, aspirations and behaviours of the users?

FIGURE 3.6 User research focus.

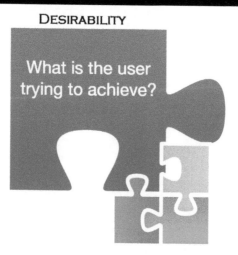

DESIRABILITY

What is the user trying to achieve?

What are they trying to achieve? What jobs are they trying to do? What outcomes are they trying to achieve? What motivates them to do these jobs and what barriers inhibit them from achieving the outcomes? How do they measure success? What emotions do they feel surrounding these jobs and their attempts to complete them?

RESEARCH IS NEVER FINISHED

An important final comment is to note that design innovation is an iterative process. When we describe here a Research phase that progresses with all its data to the next Reframe phase, we do not mean to imply that Research is finished and packaged away for the remainder of the project. By Research phase we mean a period when research activities are the dominant activities, though perhaps not exclusive. It will always be the case that the dawning of some new insight or the development of a new concept at a later stage will awaken new questions that will not be answered by the initial research activities. In such circumstances, it will be necessary to research some more and to keep doing this until a satisfactory answer is found. The same message applies to all phases of the ARRIVE process.

METHODS IN THIS CHAPTER

The methods we introduce in this Research section are our favourite and most powerful methods for human-centred design innovation research. We have created some, adapted and extended others. For most methods we provide a template designed to help you capture and present important data in a format that will be most helpful in the idea generation and concept development phases that lie ahead.

The research methods we illustrate later in this chapter are:

- Design Ethnography
- Experience (Journey) Mapping
- Empathy Mapping
- 5 Whys
- Think Aloud protocol

OUTPUTS AND TEMPLATES

At the end of the Research phase you will have a lot of data. Often, there is such a mountain of data that it is difficult to see where to start to make sense of it, how to share it and how to bring to the surface the significant themes and insights for innovation.

Making sense of the data is what happens in the next phase of the ARRIVE process: Reframe. The outputs from Research that you bring forward to do this work are your (raw) research data in the forms given below.

- **Experience maps:** An experience map or journey map is a great tool for capturing and visualising the key touch points as a user engages with a process, product or situation that you are studying. More especially, it assists representation of the user's degree of satisfaction or disappointment with the engagement and indicates opportunities for creating a more compelling offering.

- **Empathy maps:** Decisions and behaviours are strongly influenced by the frame of mind and emotional state of the subject. The emotional state is influenced by the environment where she is situated. An empathy map puts central emphasis on the person under study. What does she see, hear, think, feel, say, do in a situation, and what are the dominant positive and negative emotions that she feels?
- **Other data,** including video recording, photo images, audio recordings, notes and other collateral, which will be used for later refinement and development of insights and concepts.

Research | methods

a audit
What is known?

r research
Find out more

r reframe
Change Perspective

i ideate
New ideas & Concepts

v validate
Test with users

e execute
Develop, deploy & scale

1

DESIGN ETHNOGRAPHY
Immerse yourself in the lives of your customers. Understand values, aspirations, relationships and beliefs of those you are studying.

2

EXPERIENCE MAPPING
Experience (or Journey) Mapping helps identify, and record, high and low points in the experiences of customers trying to get a job done

3

EMPATHY MAPPING
Empathy is a cornerstone of design thinking. It helps you put yourself in the minds of customers and achieve a deeper understanding.

4

5 WHYS
Get to the root cause of problems. When you know the root cause, you are better placed to find a more complete solution.

5

THINK ALOUD PROTOCOL
A live commentary on a user's thoughts, as she performs a task, helps you to understand why she makes decisions.

In the research phase you dig deeply to understand the experiences, the interactions, the behaviours and the aspirations of your customers.

This is a method from the field of Anthropology, used for systematic study of people and cultures.

Ethnography (to write about people) is a set of practices and tools developed by Anthropologists. While traditionally associated with distant or extreme culture studies, in recent years Ethnography has become increasingly utilised in business settings. Here, it helps uncover otherwise inaccessible new understanding and insights into human behaviours, interactions and values.

Understanding people requires that a researcher immerses herself directly in the context of the lives and practices of those she seeks to know. An Anthropology adage goes, 'if you want to understand animals go to the jungle, not to the zoo'. It is not unusual for this to take months, even years.

For most managers, months of research fieldwork is not feasible so, over the years, practitioners have developed leaner approaches to Ethnography. One of the tools we find most useful for practising this 'lean' Ethnography is Spradley's nine dimensions. This framework is designed to guide the researcher as to what, when, where, why and whom to observe. Deep observation takes time to perfect, but this tool provides a structure for practising and honing the observation skills. It does not, in itself, qualify the practitioner as an anthropological ethnographer, but it brings real value to the innovator. With time and practice the novice ethnographer-innovator becomes more skilled at extracting powerful and actionable insights from ethnography.

The goal of ethnography and Spradley's nine dimensions is to immerse you in the life contexts of those you are trying to understand, to enable you to see the world through their eyes and ultimately illuminate that which is otherwise taken for granted or goes unspoken within a community.

Step by step

1. **Define what to observe**
   Keep the scope broad enough to allow you to identify what is relevant, but focused enough to be able to dig deep. For example, if you are a food manufacturer, you could observe the spaces where customers consume that type of food and look for the role that food plays in their lives.
2. **Who will you observe and in what context?**
   Whose perspective do you want to understand? For example, laboratory technicians and a process they carry out; new customers using a product for the first time; a scrub nurse during procedure. Pick a target group and context.
3. **Choose an approach**
   Pure Observation: Observe the participants as they carry out tasks and go about day-to-day activities. This observation should be unobtrusive and not obvious. You should try to be invisible.

   Intercepts: Here, you observe and interview at points of the task, to understand decisions and motivations.

Work-alongs: Here, you carry out the task alongside the user, observing and probing throughout the task.

4. **Observe**

Pay close attention to what the participant says and does not say. Look for signs of confusion, pain, pleasure, workarounds, etc. During or shortly after the observation work, you should use Spradley's nine dimensions framework to guide your detailed note capturing.

5. **Analyse findings**

Use the observations from this research combined with findings from other methods in this section to help you build a rich understanding of the people for whom you wish to innovate.

DESIGN ETHNOGRAPHY template 2/4

SPRADLEY'S 9 DIMENSIONS
Scope
SPACE
Layout of the physical setting; rooms, outdoor spaces, etc.
Actors
Names and relevant details of the people involved
Activities
The various activities of the actors
Objects
physical elements: furniture etc.
Acts
specific individual actions
Events
particular occasions, e.g. meetings, celebrations
Time
the sequence of events
Goals
what actors are attempting to accomplish
Feelings
emotions in particular contexts

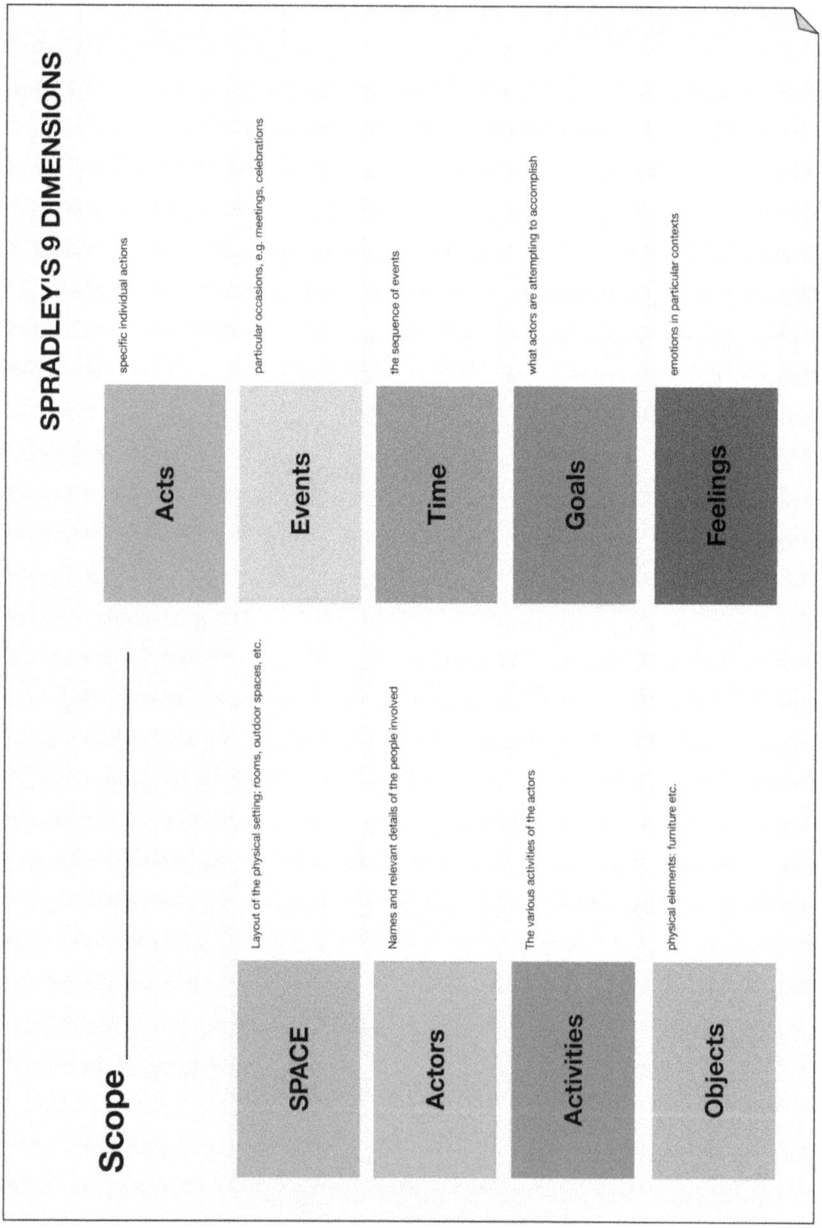

FIGURE 3.7 Design Ethnography template.

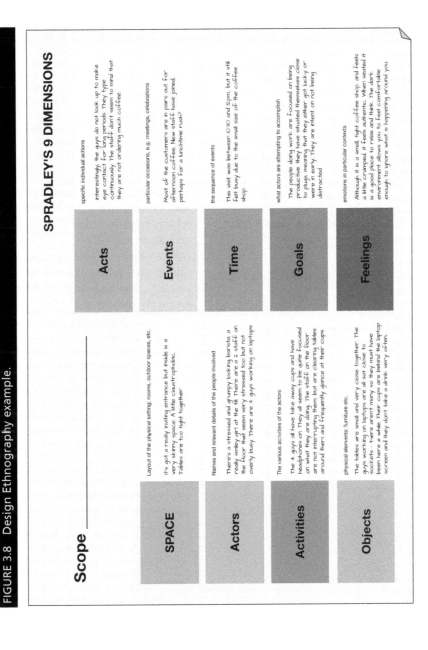

FIGURE 3.8 Design Ethnography example.

Scope

SPRADLEY'S 9 DIMENSIONS

SPACE

Layout of the physical setting: rooms, outdoor spaces, etc.

It's got a really inviting entrance but inside is a very skinny space. A little claustrophobic. Tables are too tight together

Actors

Names and relevant details of the people involved

There's a stressed and grumpy looking barista, a really smiley girl at the till. There are a 2 staff on the floor that seem very stressed too but not overly busy. There are 4 guys working on laptops

Activities

The various activities of the actors

The 4 guys all have take away cups and have headphones on. They all seem to be quite focused on what they are doing. The staff on the floor are not interrupting them but are clearing tables around them and frequently glance at their cups

Objects

physical elements: furniture etc.

The tables are small and very close together. The guys working on laptops are all sat close to sockets - there aren't many so they must have been here a while. Their cups are behind the laptop screen and they don't take a drink very often

Acts

specific individual actions

Interestingly, the guys do not look up to make eye contact for long periods. They type continuously. The staff don't seem to mind that they are not ordering much coffee

Events

particular occasions, e.g. meetings, celebrations

Most of the customers are in pairs out for afternoon coffee. New staff have joined perhaps for a lunchtime rush?

Time

the sequence of events

This visit was between 10.30 and 12pm. but it still felt busy due to the small size of the coffee shop

Goals

what actors are attempting to accomplish

The people doing work are focused on being productive. they have situated themselves close to plugs. meaning that they either got lucky or were in early. They are intent on not being distracted

Feelings

emotions in particular contexts

Although it is a small, tight coffee shop, and feels a little cramped it feels authentic. When seated it is a good place to relax and think. The dark environment allows you to feel comfortable enough to ignore what is happening around you

Q This seems wide open. How do I know where I should start?

A The key is to define something manageable, by keeping the scope clear and not too broad.
If you have customers, spend time where they shop. If it is an issue on the factory floor, go there and see how operators are behaving and interacting. Spend some time just watching, listening and noting interesting activity and behaviour. Over time your senses will sharpen.

Q How much detail should I try to capture?

A You should seek as much detail as you can in order to gain the greatest understanding possible.
Ethnographers refer to this as 'thick description'. With practice, you will get better at noticing and describing. Use full sentences for your descriptions so that you convey clear information.

Q I can't help being sceptical that this will help my innovation project! Is it not just wasting time, looking instead of doing?

A Over the past decade, many organisations have come to realise how important it is to understand their customers more deeply. Many of today's breakthrough innovations have come from better understanding customer needs, which may be unarticulated, and values, which require acknowledgement and alignment. Great innovations rarely come from technology alone, and it should not be surprising that this human-centred research should take time, as is readily accepted for science and technology R&D.
Ethnography gets you up close and personal with the consumers and users that help you shape your innovations, and who are more likely to use them because of this alignment.
Can you imagine the opposite extreme? It implies that you are so all-knowing that there is nothing your customer could contribute to your project. Such arrogance does not survive long.

EXPERIENCE MAPPING method   1/4

Also known as Journey Mapping, this is a tool that helps you distil any 'journey' into a concise, visually compelling story of the customer's experience.

Today's economy is driven by experience and, more than ever, people are aware of and react to good and bad experiences. It is essential that firms, through innovation, strive to enhance positive and eliminate negative experiences associated with their offerings. Unfortunately, many firms operate with an 'inside-out' focus as they develop products or services. They fail to consider or even see the many pains, frustrations and barriers that their customers suffer throughout the journey of finding, purchasing and using the product or service. This is a wasted opportunity, failing to innovate beyond core functionality. Often, hitherto unnoticed and unspoken details can be inspirational for creating positive, even remarkable, changes.

The experience map helps make the details of these experiences visible, understandable and therefore actionable. Practitioners using Experience Mapping know that when you can see high and low experience points on the map (and, more convincingly, repeating over multiple maps) you have valuable information to design for better experiences.

Almost anything can be 'mapped', but most frequently it is a customer interacting with a service, operating a product or experiencing a place, situation or activity. While you can map a long journey such as a 'day, or even a year, in the life', we recommend identifying a narrower focus initially and completing a number of maps.

Experience maps can be as complex as you choose to make them, but we prefer to keep them simple – breaking down the experience into no more than 5–8 steps, and keeping notes on the experience to a short, precise paragraph. Being specific demands clear prioritisation, and facilitates focus.

Step by step

1. **Define what you want to observe, and how**
   Decide the scope of the experience you will map. Identify the user participants for observation or interview. Explain the purpose of the study, and give an overview of the Experience Mapping tool.
2. **Phases of activity**
   Next, break the 'journey' down into phases of activity in the top row of the template on the next page. Do this by asking the participant to tell you in her own words how she goes about the task in question. Remind her that honesty and authenticity are essential. She must recall a specific, recent actual journey, not a hypothetical one. Keep it high level. As the participant speaks, identify the key phases being described and fill them into the map.
3. **Add detail, then plot**
   Ask the participant to walk you through the journey again, but this time with as much detail as she can remember around each phase. Encourage her to provide open and rich descriptions, where no detail is too trivial.

As she talks, record her thoughts, emotions and perceptions throughout the journey experience, good and bad. Use short sentences in the space above or below the neutral line. You can write directly on the map or use sticky notes.

Alternatively, observe the participant performing an actual journey while you note your observations and, if possible, her thoughts and emotional responses at different stages. Follow this up with a review interview as above.

Draw a line plot of the emotional experiences you have recorded at each phase.

4. **Analyse**

With the map complete, seek out additional participants and repeat the process. While you see opportunities in the highs and lows on an individual map, patterns and connections will form more starkly when you overlay a number of maps.

EXPERIENCE MAPPING template

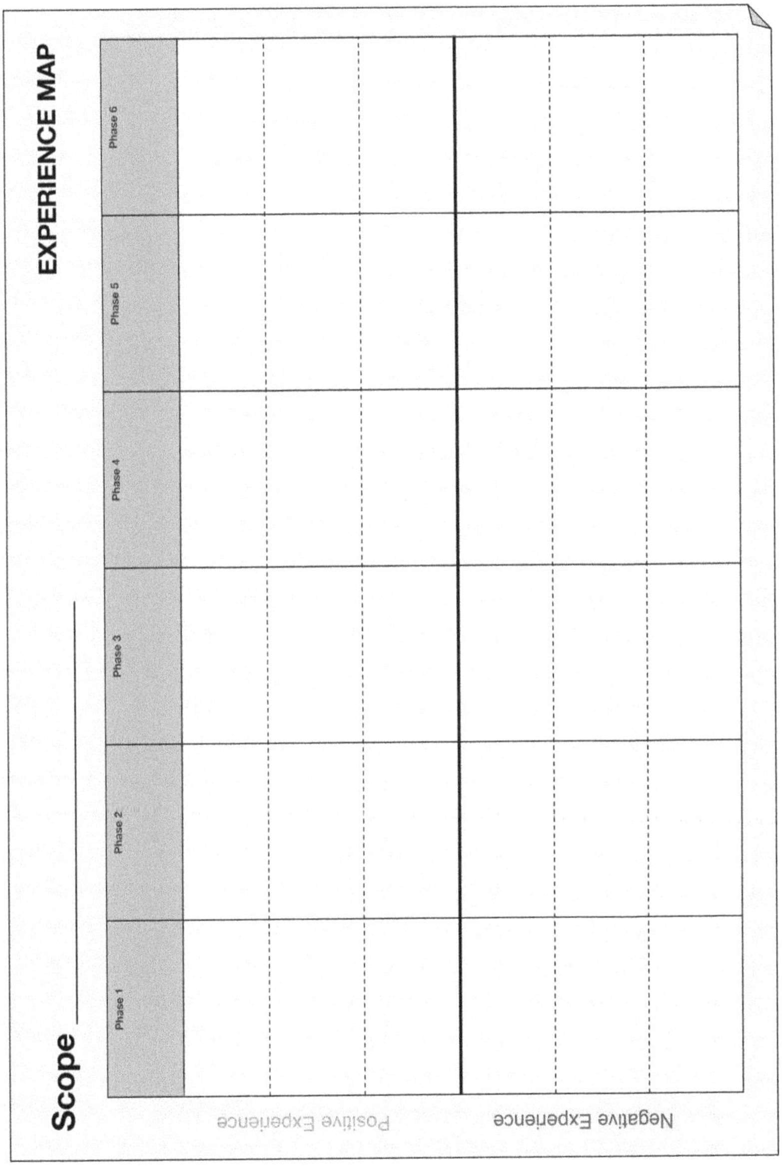

FIGURE 3.9 Experience Mapping template.

EXPERIENCE MAP

Scope

	Phase 1	Phase 2	Phase 3	Phase 4	Phase 5	Phase 6

Positive Experience

Negative Experience

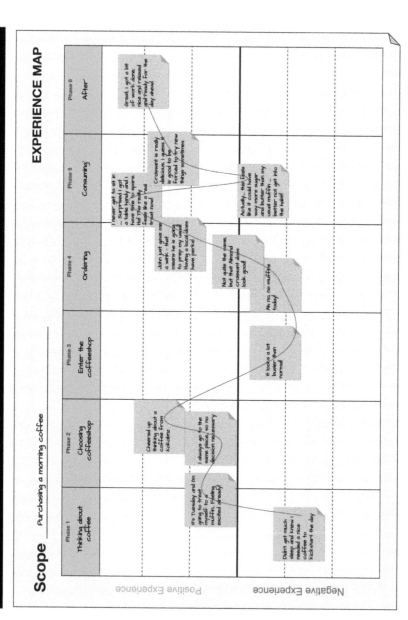

FIGURE 3.10 Experience Mapping example.

Q How much granularity should be applied in constructing the map?

A This depends on the experience being studied and your business objective. You should encourage participants to share as much as possible, but there is no need to record information you are sure is irrelevant to the project. There are 2 key objectives:

(a) to get a visual overview – hence avoid clutter of too many details.
(b) to identify key highlights – hence avoid parsing too finely, where one experience cannot be isolated from the adjacent.

If the map has over ten phases it's getting unwieldy. If it has fewer than five or six, it seems too sparse to be useful.

Q Is there anything else I should do?

A We like to see images and artefacts included with the experience map. Ideally, you should photograph a scene or activity which represents the phase of the journey. This brings the experience to life, makes it more reliably memorable and helps develop a wider empathy amongst the team for the actors in context.

Q What can I map using Experience Mapping?

A We have found anything can be mapped where a human interacts with a person, product, service or feature. It gets more difficult when you cannot see the experience in real time, and must rely on reporting. Complicated tasks with many and overlapping elements, or sensitive, personal journeys can also prove difficult. In these cases, we recommend breaking the journey down into manageable chunks.

Q Why do you say to record a specific, recent journey rather than a hypothetical one?

A Memory plays tricks, always. Our quality of recall is constantly subject to various biases, selective observations and memory deficits. The distortion is worse as the circumstances we try to recall are more distant. It is worst of all if we try to construct an imaginary scenario, based on aggregated or averaged memories. Here, imagination and fantasy too easily get mixed with fact.
In order to minimise these effects, focus on a specific, real, recent journey and consciously attempt to report on facts alone.

This is a tool that helps you put yourself or your team into another person's mind.

Empathy is a cornerstone of good design and good business. While we all empathise in our daily lives, sometimes amidst the bureaucracy of business we get separated from the complex and evolving realities of our users' and customers' lives.

The Empathy Mapping tool was developed by Dave Gray (www.xplaner. com) as a framework that allows the designer/innovator to gain a deeper understanding of a customer or service user in a specific context.

The empathy map, through dialogue, asks the innovation team to consider what the subject 'hears', 'sees', 'thinks and feels', and 'does' in her daily life. It is a simple and impactful way of seeing the world through the eyes of others, our customers and partners. It also provides a mechanism for remembering and sharing the empathic insights.

Given its benefits, constructing an empathy map is surprisingly simple. Ideally in a team setting, the mapping process allows you to discuss the circumstances of a stakeholder you wish to understand better, and to gain actionable insights by leveraging her perspective. The process of completing the empathy map will better inform and prepare you and your team to create innovative, and customer-centred solutions.

The data on the empathy map supports deeper understanding and will be a valuable source of inspiration while you analyse and synthesise your research data in the next phase of the ARRIVE process, Reframe.

Step by step

1. **Define who you want to focus on**
   First, choose your subject. You are trying to know her really well in the context and circumstances of your project. It may be a key customer or delivery partner, or another stakeholder. Decision makers are great subjects for this exercise.

2. **Observe and interview/Step into the shoes**
   First, observe the person in context, for as long as necessary and/or possible, paying attention to all the sections of the empathy map as a checklist.

   After observation, interview the subject about what you observed. Try to banish any preconceptions. Approach the exercise with an open mind and it will highlight inaccuracies and gaps in your existing knowledge as well as reveal new findings. Guide the conversation flow around all the map headings. Put on your journalist or 'Detective Columbo' hat. Do not ask, simply, "what did you hear?" Find clever ways to get the information you want in a conversational way that keeps the subject at ease and focused on the context of study rather than your questions.

   Capture your observations and interview data on Post-it notes, in short, snappy, to-the-point sentences that are ready for placing on a map later.

3. **Complete the map and analyse**

Immediately after your field research as above, work your way around the template answering the probing questions in each section with a succinct Post-it from your research notes or a newly created one to summarise your data. You may choose to do this alone, but it works very well with a team where discussion can facilitate even deeper levels of understanding. Work hard to see the world through the mind of the person you observed and interviewed.

This exercise is a powerful way to understand the subject's perspective and to record it for later use.

FIGURE 3.11 Empathy Mapping template.

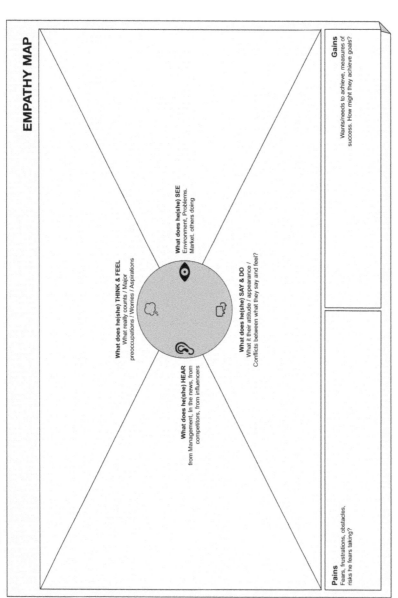

Source: Adapted from the empathy map at http://gamestorming.com/empathymap/. © 2017 Dave Gray.

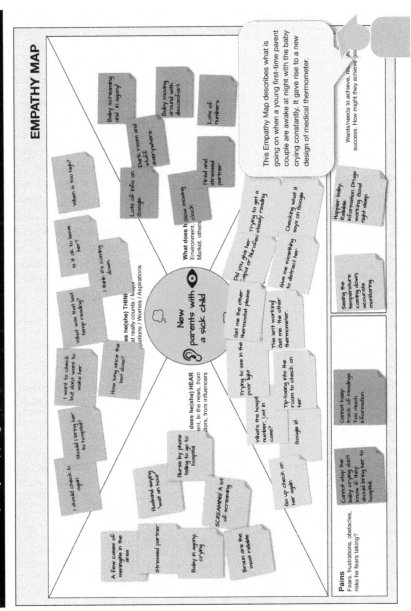

FIGURE 3.12 Empathy Mapping example.

Q It feels like I am just making up content, and that can't be right. What is wrong?

A If you're a physicist, engineer, accountant or with a similar numerate background it may seem strange, subjectively to interpret your observations. Ethnography, like many other social sciences, is not an exact, deterministic, positivist science, because the human and social world is so complex that it can never be specified exactly.
However, accepting your knowledge can't be perfect, it certainly can be better. And, you can be rigorous about making it better. We can take methodical and definite precautions, as we have alerted regularly in this book, to avoid errors and minimise distortions arising from bias, etc.
Empathy is gained cumulatively. You need some to get more. And it is never perfect, i.e. you will never know precisely what is going on in another's mind, but as you work at it more, so your intuition will improve. As you struggle to understand your target user, your enhanced empathy for her will bring great dividends in your innovation efforts.

Nothing is wrong with this. It is simply not 100 per cent definite. We highly recommend it!

Q How much detail should I include?

A During research, capture anything that seems relevant to the empathy map headings. Later, you can filter and exclude whatever seems extraneous or redundant.
Keep your observation descriptions concise and minimalist. Avoid unclear, abstract or extravagant phraseology. One or a few short sentences is usually enough to capture the detail of an experience. Do not have multiple observations in one note.

Q Should I fill the map out while talking to the person?

A The empathy map gives you a good structure. As you get more natural at it you will use the map less during interview, but there is nothing wrong with having it in front of you during your earlier experiences at interviewing.
Physically, it may be more difficult to manage this while doing field observations.

5 WHYS method   1/4

This is a method to interrogate the deeper understanding your team absorbed during research, in order to better explain a problem, behaviour or motivation.

We have all been subject to five-year-olds looking to better understand the world by continuously asking 'why'. The 5 Whys technique originated for the same reason. Asking 'why?' five times brings you closer to the root cause of a problem. When you understand the root cause of a problem, then you are better equipped to design effective and systemic countermeasures that will survive over time.

Often we are faced with problems in work and ordinary life where we are too busy, too impatient or the situation is too urgent to spend time understanding it properly. The 5 Whys technique forces the solver to spend time with the problem. As the Toyota Lean Manufacturing engineers, who originated the technique, used to say: "go and see the problem".

In the Lean Manufacturing process, the 5 Whys was created to help engineers diagnose manufacturing issues. Its simplicity and usefulness has found it applied to many other types of problems including innovation research, as we discuss here.

In its original use in manufacturing, 5 Whys is an iterative interrogative technique that identifies cause-and-effect relationships underlying specific problems. The process usually converges on a single dominant or root cause.

In our case, we use it as a methodology to uncover deeper or systemic causes of interesting user behaviours and motivations that we are interested to change.

Instead of one specific cause, there may be multiple causes associated with different user types. These are all valuable and they will be synthesised with other data in the Reframe phase (next chapter) to divine new insights and user needs.

Step by step

1. **Choose your topic and team**
   After your main body of ethnographic and other research has been done, you will have many unexplained observations. Happily, the team is now rich with empathy and broad understanding, so it is opportune to mine this for better, more profound and systemic explanations of your chosen topics.

   Choose one or more topics that you wish to study in a session. Gather your team or those members most familiar with the situations. If possible, include some users.

2. **Define the problem**
   Articulate the problem in a short, concise and detailed statement. Write it in large text on a whiteboard, leaving plenty of room for answers. Experiment with how you frame the problem. Perhaps, try more than one approach to phrasing it as a statement or question. See, the example over the page.

3. **Ask 'Why?'**

   Discuss as a group, and/or question a customer about the next direct causal factor behind the problem. It is important to ground this in factual information as much as possible. Avoid unfounded speculation.

4. **Ask more 'Whys'**

   Continue exploring deeper causes. There is no fixed rule that it must be 5 Whys, but this is usually sufficient. You have gone far enough when your team judges that you have reached a cause that, if resolved, will have a lasting and systemic effect.

5. **Countermeasures**

   List the uncovered root causes and main challenges that arise in solving them. The root cause explanations and challenges are included in later data synthesis and reframing exercises, in the next chapter.

5 WHYS template

FIGURE 3.13   5 Whys template.

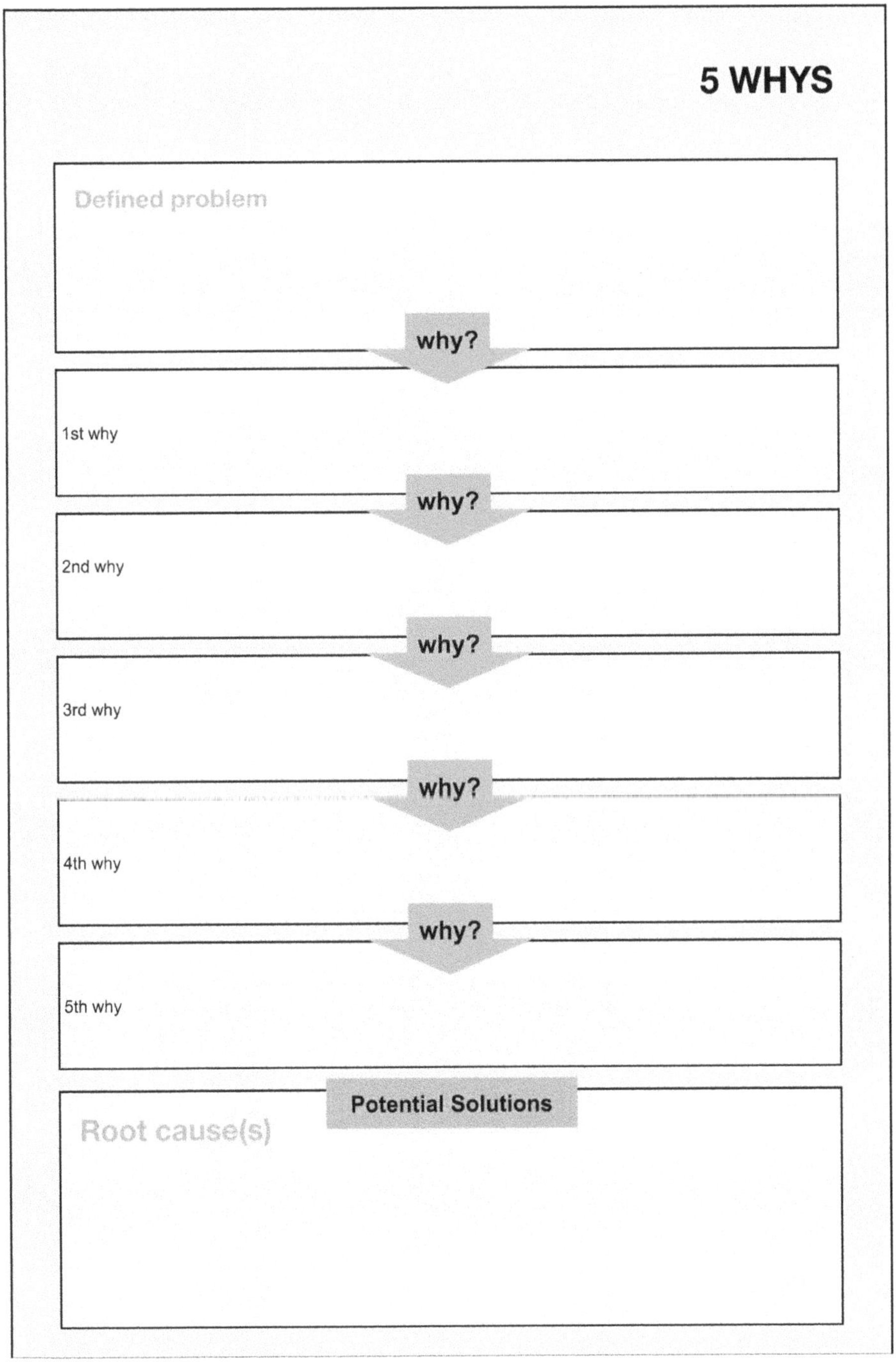

5 WHYS

Defined problem

why?

1st why

why?

2nd why

why?

3rd why

why?

4th why

why?

5th why

Potential Solutions

Root cause(s)

FIGURE 3.14 5 Whys example.

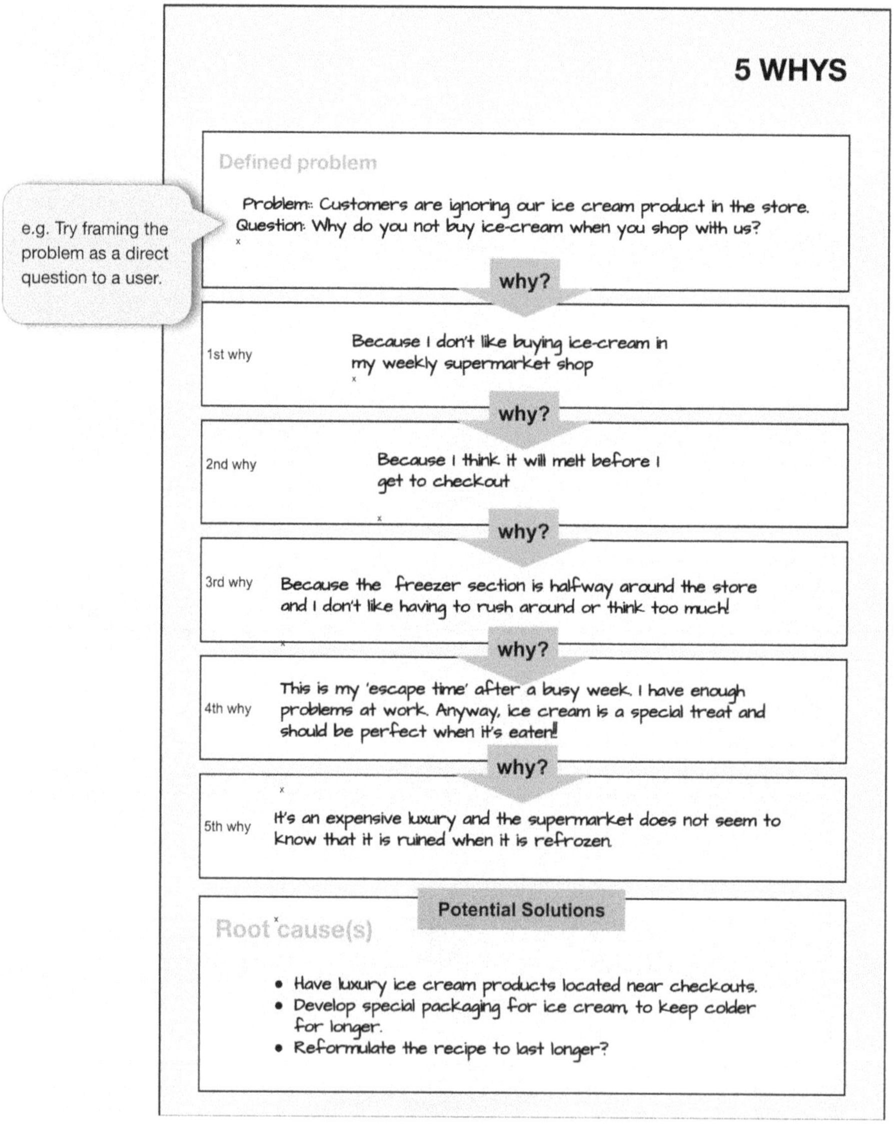

5 WHYS

Defined problem

Problem: Customers are ignoring our ice cream product in the store.
Question: Why do you not buy ice-cream when you shop with us?

> e.g. Try framing the problem as a direct question to a user.

why?

1st why — Because I don't like buying ice-cream in my weekly supermarket shop

why?

2nd why — Because I think it will melt before I get to checkout

why?

3rd why — Because the freezer section is halfway around the store and I don't like having to rush around or think too much!

why?

4th why — This is my 'escape time' after a busy week. I have enough problems at work. Anyway, ice cream is a special treat and should be perfect when it's eaten!

why?

5th why — It's an expensive luxury and the supermarket does not seem to know that it is ruined when it is refrozen.

Root cause(s)

Potential Solutions

- Have luxury ice cream products located near checkouts.
- Develop special packaging for ice cream to keep colder for longer.
- Reformulate the recipe to last longer?

Q Is there always just one root cause?

A No, there may be more than one important cause. In principle, all causes may be unified at some 'system design' level, but often that level is inaccessible to the innovator. For example, you might conclude a hospital's problems are ultimately caused by its shortage of doctors, but in a national system with limited health budget and doctor shortages, this is a moot, and not so useful, point.
So, it is legitimate to consider more than one deeper cause. But, do not allow these to proliferate easily. It quickly becomes unmanageable and therefore of little value. Try to deal only with 'root causes' that are really important and effectively independent of each other. With complex situations and multiple causes you may end up with a tree-like structure, tracking to multiple deeper causes. This is possible to manage, but more difficult.
For the reasons above, 5 Whys is most useful in more straightforward problem cases.

Q When is a good time to use the 5 Whys method?

A As suggested by their name, root causes are often hidden. When you suspect there is something deeper or hidden that might explain a puzzling phenomenon, then the 5 Whys technique is a useful way to systematically troubleshoot and improve your understanding. In addition to the root cause you learn about the process or mechanism that leads to the effect.

Q How do I know where to stop?

A This is difficult to answer, as it will depend on the complexity of the challenge and the 'track' you took in questioning it. You know you are close when asking 'why' produces no more useful responses. This may come on the 3rd, 4th or 5th why; 5 is really just a nominal number.

This is a real-time commentary, describing the commentator's own thoughts, decisions and rationale as she performs a task.

The Think Aloud protocol is a method that allows the innovator to understand the thought process of a subject as she uses a product or service, or executes a predefined task. The innovator observes and listens while the subject attempts to complete the task.

The Think Aloud method has been widely used in design, psychology and other social science contexts for many decades. There is evidence that the Think Aloud protocol provides better quality data than other methods that rely on recalling information or verbalising hypothetical scenarios, which are subject to selective recall bias. The Think Aloud protocol subjects merely report verbally, in real time, on their thoughts, without interpretation or further processing. While there may be some cognitive load associated with the verbalisation, it is minimal and usually the subject quickly becomes accustomed to it.

For this reason, concurrent (real-time) Think Aloud protocol is most valuable. However, if necessary, it may be performed retrospectively, whereby the user recalls as much detail as possible, soon after the task is completed.

The method is valuable in identifying and bringing light to areas where the user is confused, has difficulty or is wrongly directed. It helps the researcher understand thought processes and assumptions made by the user.

Think Aloud is best performed with one or two observers; more may be distracting to the subject. Care must be taken not to cause the user to be distracted or self-conscious, which would result in poor data from non-natural thinking and reporting.

Step by step

1. **Identify the task**
   The task should be scoped such that it does not take too long to complete and has a clear starting point and objective.
2. **Prepare the subject**
   Explain the task that you want the subject to complete. Explain the protocol and reassure the subject that it is the process not the subject herself that is being assessed. Remind her that she can stop the task at any time, and that she may ask questions if she wishes to.
3. **Begin the task**
   Remind the subject that the only time you will interrupt is if she goes silent and forgets to 'think aloud'. If this happens tell her you are going to say, "what are you thinking?" to remind her.
   When the subject is ready and has no more questions you may begin the task.
   As the subject is talking you should take detailed notes, jotting down as much as possible and noting the sequence of the various thoughts, actions and action stages. Everything is important. You may

find it helpful to record the notes live on an experience map. Additionally, with permission, it is very helpful to video- or audio-record the activity.

4. **Debrief and close**

When the Think Aloud is complete, you should now guide the participant again through the task and her commentary. Ask her to elaborate on certain aspects that are unclear, particularly interesting or puzzling for you. For example, there may be stages at which she reported confusion where you did not expect it: what was the cause of the confusion? Or, she may have taken a shortcut that you didn't expect: what led her to recognise and try that?

FIGURE 3.15 Think Aloud protocol typical set-up.

FIGURE 3.16   Think Aloud protocol transcript example.

Notes: Think aloud - Ordering a coffee

"I'm thinking this queue is a little long but should move fast, as starbucks staff are very efficient"

"I'd better have a look at the menu"

"I always go for the same thing but I like to check the price every time so I can get my change ready"

"A slightly longer queue is nice as it gives me time to scan the menu and maybe have a look at the cakes"

Now what?

"… I just can't remember if I locked the office door!"

"It's nearly lunchtime and I am wondering should I order a cake or a sandwich"

"The last time I was here the sandwich wasn't great - I could have been unlucky but it has put me off. I've only a limited amount of calorie credits and I don't want to waste them!

"I'm going to go for one of those brownies I think…..but that cookie does look good!"

"Yup, brownie…...It is a bit big though…..I wish they were smaller"

Now what?

"€3.70 for a coffee is crazy, the place down the road is actually cheaper and nicer" "I always think that when I am in the queue here, and it's too late" - "I guess it's just convenience!"

"Hmmm, a cappuccino would be nice...no, i'll stick with americano...I heard an Italian chef once laugh at people that order cappuccino during the day" - "I've been paranoid ever since!"

"Hi, can I have a regular americano pease"
"Can I have it here, but in a takeaway cup"

" I don't know why, but I just prefer a takeaway cup. It just feels like you get more, it stays warmer longer and you know, you can get up and leave if you have to"

"I both love and hate when they ask for my name….It just feels a little fake, but then when they call you by name it's kind of nice!"

Q Do I really need the subject to be carrying out the task? Could I not simply ask them to tell me about the last time they did it?

A No (to the second question). When recalling events we often leave out important details, due to forgetfulness or selective recall bias. These details, no matter how small, could hold important clues especially when looking for levers to innovate and improve.

With concurrent Think Aloud, the participant goes into the details of what they are experiencing there and then, how they are making decisions and what factors are driving those decisions. Ideally it becomes a stream of consciousness with no filtering.

If you can achieve the latter, it's the best data you'll get.

Q What size task should I choose for the think aloud activity?

A We recommend selecting a relatively tightly constrained task. This is because thinking aloud itself imposes cognitive load, and over time may reduce the quality of the information and annoy the participant. It may be worth breaking large tasks into multiple tasks and providing a break in between.

Q How should I capture the information?

A Wherever possible, we prefer to record the activity on video, especially when it involves doing a tangible task, e.g. finding and ordering something online. This allows us to capture all the data of expressed thoughts and actions, for later analysis.

There is great value in paying close attention to the subject in real time, without distraction of notetaking, so that subtle clues of confusion, hesitation, satisfaction or other unexpressed thoughts may be noted.

BIBLIOGRAPHY

Klein, Gary (2013). *Seeing What Others Don't: The Remarkable Ways We Gain Insights*. Nicholas Brealey.

Spradley, James (1980). *Participant Observation*. Cengage Learning Inc.

The Economist (2008, November). The Hawthorne Effect.

Reframe

The problem framing defines the solution space.

The reframe phase seeks to interpret your research data and to guide the rest of the project based on that interpretation.

By reflecting collaboratively and intensively on the wealth of data from your team's research activities, you will uncover and crystallise patterns, insights, user needs and desired outcomes for targeted users. The needs and desired outcomes define the basic solution requirements. The unexpected, hence special, insights inspire exceptional and differentiating solution possibilities.

With a more comprehensive understanding, now you can reframe your approach to the problem with a fresh perspective. Choose a frame where greatest potential exists for transformative impact, aligned with your firm's capabilities, and which will guide the innovation project to completion.

FIGURE 4.1   Chapter 4 outline.

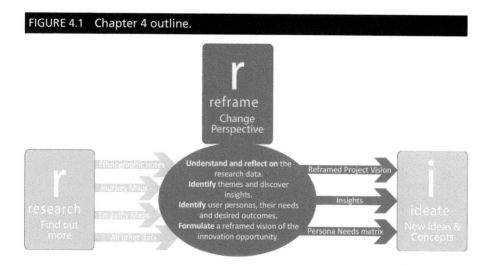

THE NATURE OF LEARNING AND CHANGING PERSPECTIVE

We begin this chapter by touching in a lightly philosophical way on supporting background for the activities that are to come. What do we understand by knowledge, learning and perspective?

Here are two axioms:

- Nobody has the whole truth!
- We construct our version of reality from our own experiences and beliefs.

Remember the fable of the elephant and the blind men? A king in ancient India brought six blind men to an elephant and asked them to describe it. The blind men touched different parts of the elephant and declared the elephant is like … a spear (tusk), a snake (trunk), a wall (side), a rope (tail), a fan (ear), a tree (knee). Each of the blind men was partially right and yet all were wrong. Each of them had only partial understanding because of his incomplete experience of the elephant.

At the start of an innovation exploration you, also, and your team are blind to the full picture, being unfamiliar with many aspects of the total problem space and bringing with you limited perspective with cognitive biases that are unavoidable. Now, after research, you have a richer understanding and can see different perspectives that help you envision how the problem space might be transformed.

But it is not always straightforward. Usually the research process brings out so much data, with varying quality, that it is hard to separate

FIGURE 4.2   The blind men and the elephant.

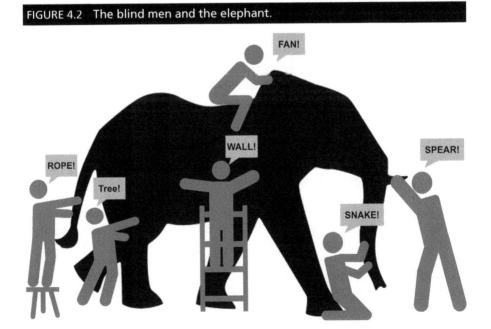

signal from noise. The challenging task is to uncover and isolate what may be significant and useful. What is the target user really trying to achieve in the context? What is the priority ranking of the users' objectives? Is there hidden within the data a key insight or two that might unlock opportunity to make exceptional impact on the problem space? There is, often! Is there a perspective on the problem that helps you formulate a vision for transformation? While it may be challenging, it is also thoroughly exciting and fulfilling.

In order to interpret qualitative data properly, you have to accept the fragile and imperfect structure of your prior knowledge and assumptions about the area under study. After all, if you believed you already had a robust understanding, why bother with research? There are two concepts about the nature of beliefs and learning that have been developed by the father of organisational learning, Chris Argyris, which help us understand how to go about productive reframing (Argyris, 1983, 2004).

The **Ladder of Inference** helps us understand how to get to a **broader framing** or alternative-perspective understanding.

The model of **Double Loop Learning** helps us understand how to get to a **deeper framing** or root-cause understanding.

These are, respectively, analogous to the pan and zoom functions of a camera. See Figure 4.3.

Let us look at these in a bit more detail, in order to understand the nature of reframing in practice.

LADDER OF INFERENCE

It rarely occurs that what you believe to be true is in fact a complete truth. It is hard to find a situation that cannot be given some fresh meaning from another perspective. Even well-meaning eyewitnesses to a car accident inevitably give police different versions of events, due to their varying observational skills, alertness, recall and bias, whether those biases are conscious or not. Firms, too, view markets in a certain way. They use certain time-honoured and HQ-directed conventions to frame their competitive set to the extent that often, in multinational organisations, local subsidiaries are not able or allowed to absorb richer local data into the corporate knowledge base.

Ask yourself, was there ever a conversation you have had with someone you respect where you did not come away with fresh information or at least a more nuanced understanding of the topic? The honest answer must be rarely or never.

According to Chris Argyris (1983), what appears obvious to us is really the set of assumptions that we believe to be true. In turn, these assumptions are our interpretations of data that we filter selectively according to what we consider most personally interesting or most relevant. Of course, we base what we consider relevant on our beliefs, and so the data we choose serves to reinforce our beliefs and complete the bubble of self-sealing logic. This model of moving from experience through interpretations to belief and on to action elegantly illustrates the tendency we all have to jump to

FIGURE 4.3 (a) Pan reframe and (b) zoom reframe.

(a)

(b)

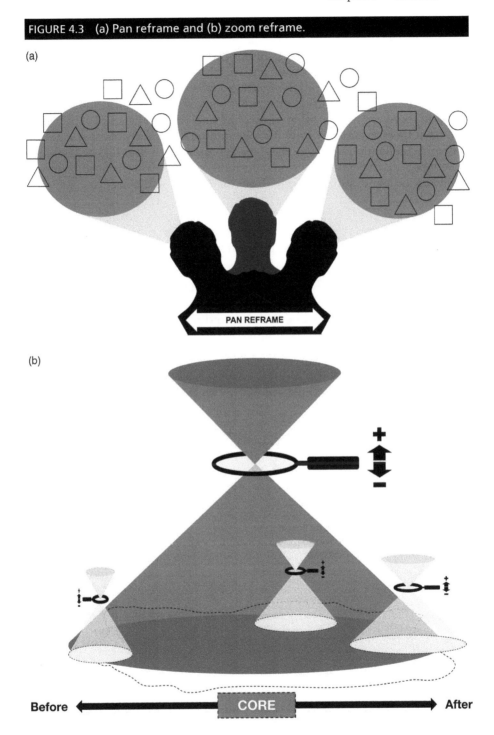

conclusions too early with limited data or blindly to accept invalid assumptions. See Figure 4.4.

In his book *Liminal Thinking*, author Dave Gray (2016) develops the concept of a Pyramid of Belief, based on Argyris's Ladder of Inference, to illustrate the way people with different backgrounds and experiences may exist in mutually independent worlds of *self-sealing logic*. Gray illustrates the way to transition between these worlds – to explore one's own beliefs and experiences and from there to consider how the experiences and beliefs of others may be different. See Figure 4.5. Obviously, for an innovator individual or team, the experiences and beliefs of users and customers of their products and services will likely be significantly different from the innovators themselves.

When you look at a situation from a different perspective you uncover new meanings and perceptions of reality, and with these new beliefs you can be inspired to act in new ways, with a broader framing. Applying this model to the process of innovation leads to two imperatives:

1. You must trawl broadly for relevant data. This is the role of the Audit and Research phases, which we have studied in the previous two chapters.
2. You must carefully consider, interpret and re-interpret all data, and especially new data, to seek new understanding. This is the Reframe phase, which is the subject of this chapter.

FIGURE 4.4 The Ladder of Inference.

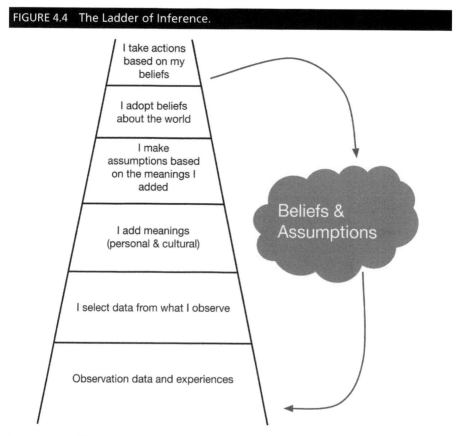

Source: Argyris, 1983.

FIGURE 4.5   Transitioning between belief systems.

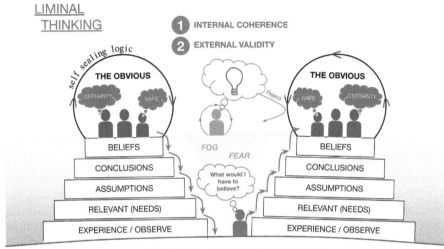

Source: Gray, 2016.

DOUBLE LOOP LEARNING

A second beautiful concept from Chris Argyris (with Donald Schön, 1978) helps us get to a deeper, root-cause reframing.

The model of Double Loop Learning describes how a good learning organisation achieves continuous improvement by constant reflective questioning of the 'governing variables' that determine its actions and their consequences. By questioning and revising the underlying governing variables – goals, values, plans and rules – of a situation in the light of already implemented actions and

FIGURE 4.6   Double loop learning.

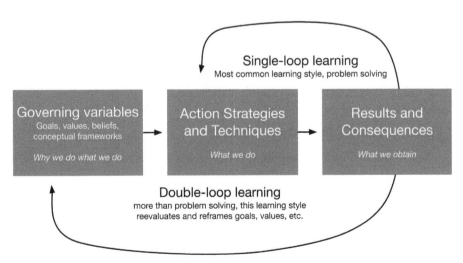

Source: Argyris and Schön, 1978.

their consequences, we may come to revise these and hence change players' courses of action and achieve improved consequences or outcomes.

Single Loop Learning implies the ability to detect and correct errors in certain operating procedures, whereas Double Loop Learning implies being able to see beyond the situation and query operating norms. This has many analogues throughout innovation and quality practice, such as root-cause analysis or the 5 Whys of lean manufacturing (see Chapter 3). The model of Double Loop Learning leads to deeper, more holistic framing.

IDENTIFYING THEMES: SYNTHESISING NEEDS AND INSIGHTS

How do you synthesise relevant themes, user needs and new insights from raw research data? How do you get clarity from the haze? Before starting, you must recognise the synthesis process requires intensive focus and collaboration because it is difficult and mentally demanding. The research team must ensure that it sets itself up under the right physical, social and psychological conditions that are conducive to productive and creative work. See Kolko (2011), which we discuss in more detail below.

Later in this chapter, as usual throughout this book, we will introduce you to details of our preferred methods for synthesis.

When we talk about synthesising themes from research data, we mean identifying patterns or groups of data points that form a coherent cluster of observations relating to distinct user needs, aspirations, motivational drivers or behavioural patterns. To achieve this requires an intuitive act of pattern recognition by the innovation team and demands focused attention, as we discuss below.

Some themes serve to describe particular needs and aspirations (or desired outcomes), which constitute requirements for any solutions that may be developed. Needs and aspirations answer the question: *What is the user trying to achieve?* Sometimes this is referred to as the user's job-to-be-done.

Other themes shed light on the users' motivational drivers, behaviours and circumstances, the *why* behind the *what*. They provide a deeper understanding and clearer mental model of what is important to achieving the desired outcomes successfully. We refer to this deeper understanding of a particular context as an insight. Insights inspire and support finding creatively differentiated solutions that will delight and be especially compelling for the user. Some themes combine all of the above.

PERSONAS AND PERSONA NEEDS

In the previous chapter (chapter 3, Research) we discussed how you might identify and segment your prospective users, and we introduced the valuable concept of personas, which you construct from your research data. Each persona represents a type or category of key user. We describe a simple method for constructing personas later, in the methods section of this chapter.

Your research has identified personas that have needs, or desired outcomes. For each need you must define exactly what each persona with that need is trying to achieve within the context and scope under study, together with his criteria for success, i.e. what metric will each persona use to measure satisfaction with an outcome, and what level of the metric constitutes success. Of course, different personas will have different priorities here.

Personas: pro or con?

Some authors assert the use of personas is misguided and not fruitful. Usually, such critics assume that personas are a priori products of the innovating team's own imaginations, superficially constructed to *guide* the research process, rather than being an output of research. We agree with the criticisms of personas used this way. But, these are not personas as we use them here.

Let us emphasise here, the design innovation persona is not a figment of the imagination, nor is it an average, perhaps unrealistic summary of a broad group of people. It would be difficult to empathise with this. Further, the research-based nature of design innovation personas ensures they are not one-off exceptions.

Before going on to ideation and concept development – and we know you will be champing at the bit to do so quickly – it is important to record your clear understanding of these needs. You do not need to list them all, but certainly you should document the ones you believe are central to the user(s) in the context under study.

If you or your organisation are quantitatively inclined, you may employ a scoring system to grade the importance and potential added value of satisfying each need for the different personas. Thus, a prioritisation of needs is possible by numerical ranking of the results. You may even wish to give individual personas different weightings to represent their relative importance to your target business area. See Figure 4.7 (without persona weightings for clarity). Many design-oriented organisations prefer, however, to retain the richness of data in qualitative form rather than reduce to the starkness of numerical grading at this stage. In any event, your team's desired outcome is the same, viz. the innovation objective is to find the particular needs that are important for most personas and where a successful resolution will resonate best with their aspirations.

Therefore, in addition to our Reframed Project Vision and Insight Notes (see later in this chapter), we also document the principal needs and desired outcomes that were uncovered in the research – functional, social or emotional – with desired outcomes and outcome metrics. This *Persona Needs Matrix* forms one of the key evaluating and shortlisting tools for concepts developed through the ideation phase described in chapter 5,.

We illustrate the use of the *Persona Needs Matrix* template at the end of this chapter.

Figure 4.7 shows an example of user needs found in researching showering experience for a domestic shower provider. A UK design firm found

FIGURE 4.7 Persona Needs Matrix example: domestic shower enclosures.

Persona Needs Matrix
Summarise outcomes your personas are trying to achieve. Add needs until all are identified.

Shower enclosure study — June 2019

Theme *What is the high-level need category?*	# *Need id*	Need *What 'desired outcome' is the user trying to achieve?*	Success metric *How will the user measure success?*	scores: Persona 1	Persona 2	Persona 3	Sum of scores	Rank
Physical shower environment	1	Comfortable, spacious in-shower space	Floor area; Shower height; Shower head height	3	3	1	7	2
	2	Comfortable, safe access	Step height; Door width	3	1	0	4	10
	3	Ergonomically & aeshetically available showering accessories	Tidiness and convenience: Soap bar, shampoo bottles x 3, loofah; nothing on the floor	3	2	0	5	8
User wellbeing	4	Feeling refreshed	Good mood; ready frp the day!	3	3	3	9	1
	5	Relaxation and de-stressing	No in-shower stress factors: e.g. water pressure, water temperature, steam build-up. Pleasant and positive aesthetics	2	2	-2	2	11
Maintenance	6	easy weekly cleaning	5-min clean; low levels of scale build up, drain blockage, dirt trap corners	3	2	2	7	2
	7	no necessity for daily cleaning	simply walk away, and it appears clean	2	2	2	6	4
Purchase and installation	8	Style and colours to match bathroom design	Wide range of contemporary and classical colours and textures	2	3	0	5	8
	9	Short delivery time	Days to installation (max 5)	1	2	3	6	4
Installation	10	Quick one-step installation	Under 3 hours; no dirt; fully sealed	1	2	3	6	4
	11	DIY installation is easy	Clear no-confusion instructions. Back up help available.	2	2	2	6	4

-2 = useless clutter; 0 = indifferent; 1 = nice-to-have; 2 = very desirable; 3 = essential

the following themes (among others) to be of particular concern to various end-users.

- Physical shower environment (especially, concerns about slipping in the shower)
- User well-being
- Product maintenance and cleaning
- Purchase and installation

A quantitative ranking is indicated in Figure 4.7. The scoring system used is detailed in Table 4.1.

Remember, this matrix is a summary dashboard of core findings in respect of needs and desired outcomes, from the Research phase. Its purpose is to be a tool for thinking and prioritisation, and a reminder checklist. It is not meant to replace the richer and more inspiring *Insights* and *Reframed Project Vision*, described below. Rather, it complements them.

Additionally, the supporting statistics and rich user quotes and observation notes (from the Audit and Research phases) should accompany the Persona Needs Matrix so as to retain the deeper understanding of its genesis.

Table 4.1 Scoring scheme for ranking persona needs.

3	= essential
2	= really wants it
1	= nice to have
0	= indifferent
−2	= negative distraction/in the way (clutter)

Remember, also, that the Persona Needs Matrix captures needs, which have not yet been converted into features. Indeed, you may have no idea yet how to achieve satisfaction of the needs; but, that is the activity for the Ideation phase that comes next.

Between granularity and inspiration

Some practitioners, including many champions of jobs-to-be-done methodology, create an elaborate, closely specified and exhaustive table of Core Jobs, Related Jobs, Consumption Chain Jobs, Emotional Jobs and Social Jobs, each with the associated job statement and desired outcome statement. The result may be, and often is, hundreds of such statements.

We tend to shy away from such attempts at fine-grained precision. We believe that important intuitive and creative opportunities for delivering extra satisfaction to the user may be diluted or lost in the granularity of attention, long time and intensive effort required to do this.

Balance is needed. A positivistic, granular, quantitative approach is valuable for its clarity and definitiveness, but it can easily flatter to deceive that it is somehow more scientific, simply because of the quantity of fine detail displayed. Alternatively, an interpretivist, qualitative, human-empathic approach can support a special type of insight-driven creative inspiration; yet, there can be a danger of fantasy losing touch with fact.

RESEARCH INSIGHTS

In the last chapter, we introduced the definition of an insight by Gary Klein as an unexpected shift in understanding or a shift to a better story or mental model, where the new mental model is more accurate, more comprehensive, more useful and perhaps surprising. According to Klein (2013), insights are triggered by observations of contradictions, connections, coincidences or curiosities, and by circumstances of creative desperation. In addition, insight generation is enhanced by having a mindset that is prepared to challenge flawed assumptions.

Later in this chapter we show you how to generate insights from raw research data. We advise you to elaborate on and record a select few insights in the form of Key Insight Notes (Figure 4.8) in the form described below. In the next chapter, Ideate, we provide further guidance on converting insights and needs into How Might We statements, which fuel the ideation process.

INTENSE DATA SYNTHESIS NEEDS A 'FLOW' EXPERIENCE

Analysing data and synthesising insights is a creative process. In his excellent book about design synthesis, Jon Kolko (2011) remarks that effective synthesis operates best with an experience of flow. This is the term coined by psychologist Mihaly Csikszentmihaly (1990) to describe when an individual operates with complete attention to the activity in hand and is highly effective in the task. Flow arises when one is challenged to the edge of, but not beyond, one's

FIGURE 4.8   Key Insight Note example.

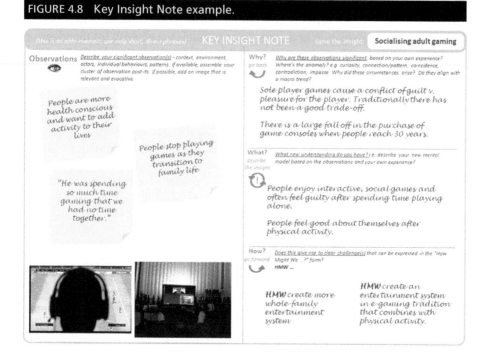

capabilities, while experiencing a state of consciousness that is "automatic, effortless, yet highly focused".

Kolko elaborates on Czikszentmihaly's four attributes of flow and relates their relevance to data synthesis. These are:

- **There is immediate feedback to one's actions.**
  Immediate feedback is achieved in data synthesis by working collaboratively and visually. Visualising the data in clusters while being able to dynamically re-arrange them helps creative thought and communication. Getting immediate reactions from colleagues heightens the sensitivity and awareness of implications of each new thought that is enacted in a visual 'move'.
- **Action and awareness are merged.**
  Maintaining the sense of flow in synthesis requires dedicated space, with no distractions and complete attention by all to the job in hand. The space must facilitate isolation of the team and enable members to focus solely on the data. Phones, email and all external distraction must be barred for the few hours' duration of the synthesis exercise. Equally, there must be no obstacle to thought imposed by lack of skill with a particular tool. This applies equally for all team members.
- **There is no worry of failure.**
  As Kolko (2011) says, "constant introspection related to failure can lead to a hyperawareness of technique [... which can] completely halt progress". The innovator needs to trust that synthesis will occur when the appropriate methods are used. This frees the mind to explore and follow new intuitions productively and without unnecessary caution.

- **Self-consciousness disappears.**
  The process, actions and insights take over and protection of ego and bias is relegated to insignificance in favour of exploring new perspectives on the data and learning from each exploration.

When synthesising data to extract insights and form a *Reframed Project Vision*, the really successful innovator teams achieve a state of flow where the team glides, as it were, around the data in a coordinated dynamic dance, forming and reforming the emergent themes and creating new meaning from what was once a dishevelled aggregation of disparate data points.

PROBLEM FRAMING DEFINES THE SOLUTION SPACE (NEW FRAMES LEAD TO NEW SOLUTIONS)

Say you have studied and dissected the research data. You have identified many needs and desired outcomes. You have extracted some keen insights from the data, which you are confident you can leverage to support a compelling innovation offering. Now you need a vision, because every good project is led by a guiding vision. The vision is a unifying target and motivating North Star for every team member throughout all of the solution-generating and execution stages that are to come. Maybe you already had a vision to start with but, with the benefit of your research, it is likely that it will have changed in substantial ways and, at least, you can now be more specific and clearer in expressing it.

A vision exists within a frame. The frame represents a paradigm or worldview where a particular pattern of relationships is presumed to apply. As with metaphors, the presumed patterns may have greater or less validity, but they always provide a focus and a creative sandbox for innovative effort. In the words of Kees Dorst, author of *Framing Innovation* (Dorst, 2015), a frame is "an organisational principle or a coherent set of statements that are useful to think with". Put another way, a frame is "the proposal through which, by applying a particular pattern of relationships, we can create a desired outcome [to complex problem solving]".

By reframing, we mean taking a new perspective on the problem area. That is, we presume a new organised set of relationships, the adoption of which is informed by new interpretations of data, in the form of insights. Usually, the new frame is inspired by a single outstanding insight or integration of multiple insights. Often, it may be represented by metaphor, as a means to succinctly define the set of relationships.

Reframing

... presuming a new organised set of relationships, the adoption of which is informed by new interpretations of data, usually in the form of insights.

A reframe is the transformative outcome that comes from reflections on insights, and is an intuitive and creative act. With design innovation methods, the insights come from interpretations of newly acquired data, especially the

FIGURE 4.9   The Reframe Vision is pivotal.

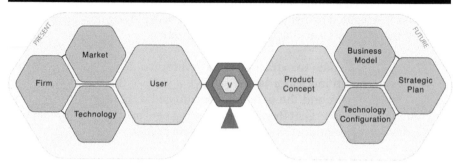

rich qualitative data of ethnographic-type research. Designers and the majority of great innovators work with this data by collaboratively sifting through it, repeatedly re-organising it, looking for patterns and inferring causality, which in turn inspires new frames within which the innovation team may generate new ideas and hypothesise new solutions.

Available solutions are different for different frames, which leads to the conclusion that *the problem framing defines the solution space*. In the ARRIVE process of Design Innovation, the framing of the problem and the high-level scope of the solution space are represented through the *Reframed Project Vision*, which becomes the guiding beacon for subsequent idea generation. Clearly, the *Reframed Project Vision* is pivotal in linking research to future solutions. Choosing and applying a frame for solutions is the most strategic and defining stage in the innovation process.

KINGS CROSS REFRAME

As an example of reframing, in his book *Framing Innovation*, Dorst (2015) describes the Kings Cross (Sydney, Australia) *Designing Out Crime* project. Kings Cross is a late-night entertainment district in the City of Sydney, which had a reputation for drunkenness, petty theft and sporadic violence. In the later hours of night-time, the clubs close and drunken patrons are disgorged onto the street. As the crowds thicken the problems worsen and sometimes people get seriously hurt. The city authorities' classic response to this is to increase the security profile with greater police presence, CCTV cameras, more club security personnel and related efforts. However, despite increased police arrests, public safety did not seem to be improving.

Designers from the University of Sydney's *Designing Out Crime* research centre reframed the problem. From their research, the designers identified a key insight that the people who get into trouble are mainly young people wanting to have a good time and are not usually criminals out for trouble. They further recognised the key problem was that each night over 30,000 people came onto the streets of Kings Cross as the pubs and clubs closed, and the streets had no structure or facilities to accommodate them. The problems arose from the disorganisation of the area and its inability to cater for the crowds. Drawing on their key insight, The *Designing Out Crime* group reframed the problem as a crowd organisation and entertainment issue and used the metaphor of a large music festival to help design remedies for the situation. For example, festival organisers would organise chill-out zones and offer continuous attractions to make sure that people would move around. Adequate toilet facilities would ensure people did not use the public streets once they had left a club and failed to get admission to another. Instead of high concentration of burly and imposing security guards, the designers proposed a system of very visible young guides in bright T-shirts, who help people find their way around the area and who are available when help is needed. Additionally, of course, transport into and out of the area would be radically improved particularly at the peak hours to avoid people being trapped without means of exit.

Adopting the new frame of crowd entertainment rather than security enforcement helped generate a wide range of new proposals to bring improved outcomes to the problems at Kings Cross.

REFRAMING AS CREATIVITY FACILITATOR

Note that reframing is often used by designers on a tactical level additionally as an idea generation tool. A temporary new (sub-)frame imposes constraints that facilitate creative idea generation. See examples of 'metaphors' and 'through the eyes' idea generation techniques in the next chapter.

As a side observation, the methods of qualitative data synthesis and reframing, sadly, hold a lot of mystery even for many designers, and invoke scepticism in many of the more rationalist professionals. Classically trained minds prefer conclusions that are straightforward and linear deductions from 'clear' and relevant data, where it is assumed that these can be isolated from other research outputs by applying criteria that are determined beforehand. However, strict evaluating criteria are implicitly correlated with framing and cannot be determined in advance of reframing, and too early fixing of them results in a form of circular reasoning or at least scope-limiting. It is the designer methods that most likely will lead to breakthrough innovation because they sustain for longer a multiplicity of possible framing options.

REFRAMED PROJECT VISION

Our objective in this chapter has been to de-mystify the concept and methods of reframing and show you how to go about interpreting data and forming a vision that holds great potential for your solution. Productive reframing depends on good research, an open and trained mind for new pattern recognition, and intensive insight synthesis, which results in an actionable and enlightened *Reframed Project Vision*. Note that all *Reframed Project Vision* statements refer to the key users, the context, and the functional, social and emotional outcomes to be achieved. Formulating a *Reframed Project Vision* will be described in more detail later in this chapter. Figure 4.10 shows a summary example relating to the Kings Cross problem.

VISION BY DECREE

We have emphasised above that there is always something to be learned from good research early in the project. However, circumstances do not always facilitate an ideal process. Despite this, we always recommend a period of research by the innovation team, to the greatest extent possible. This applies,

FIGURE 4.10   Reframed Project Vision (Kings Cross).

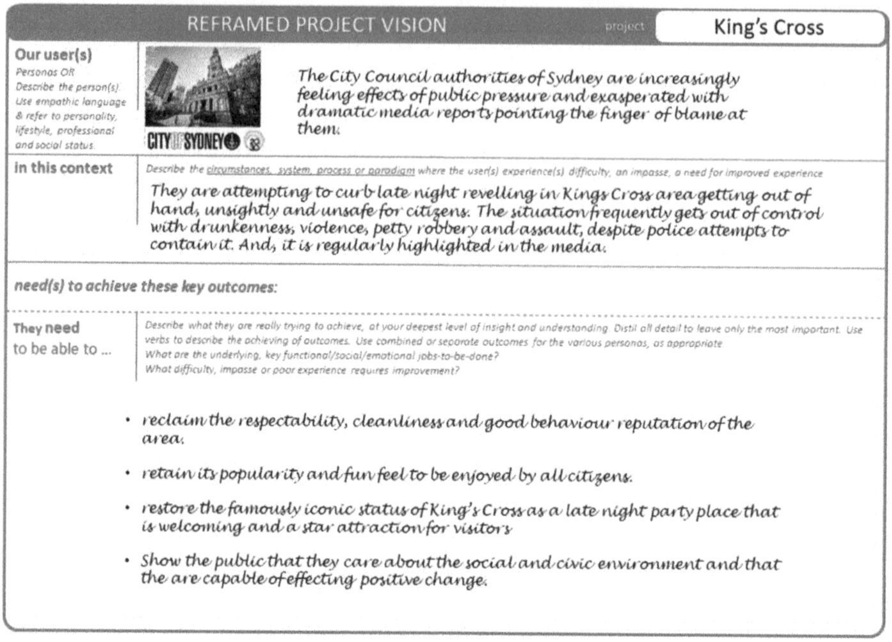

REFRAMED PROJECT VISION	project	King's Cross

Our user(s)	
Personas OR Describe the person(s) Use empathic language & refer to personality, lifestyle, professional and social status	The City Council authorities of Sydney are increasingly feeling effects of public pressure and exasperated with dramatic media reports pointing the finger of blame at them.

CITY SYDNEY

in this context   Describe the *circumstances, system, process or paradigm* where the user(s) experience(s) difficulty, an impasse, a need for improved experience

They are attempting to curb late night revelling in Kings Cross area getting out of hand, unsightly and unsafe for citizens. The situation frequently gets out of control with drunkenness, violence, petty robbery and assault, despite police attempts to contain it. And, it is regularly highlighted in the media.

need(s) to achieve these key outcomes:

They need to be able to ...

Describe what they are really trying to achieve, at your deepest level of insight and understanding. Distil all detail to leave only the most important. Use verbs to describe the achieving of outcomes. Use combined or separate outcomes for the various personas, as appropriate
What are the underlying, key functional/social/emotional jobs-to-be-done?
What difficulty, impasse or poor experience requires improvement?

- reclaim the respectability, cleanliness and good behaviour reputation of the area.

- retain its popularity and fun feel to be enjoyed by all citizens.

- restore the famously iconic status of King's Cross as a late night party place that is welcoming and a star attraction for visitors

- Show the public that they care about the social and civic environment and that the are capable of effecting positive change.

as explained below, even when the innovators already consider themselves 'expert' and have a strong vision. Alternatively, the team's project may have been commissioned by a strong boss with a fixed mind. In many practical circumstances, the strength of the organisation's leadership and the pressures of delivery deadlines do not allow for a full research programme.

Steve Blank, father of the Lean Startup methodology and author of many books and publications on the topic, has said that the vision for an entrepreneurial startup comes from within the entrepreneur, based on his own life's experience and passion. Such an entrepreneur may have had a career immersed in a particular sector or may have studied a sector in depth out of a driving passion to improve the lot of a targeted group or to add value in some respect. The driving passion can come in many forms. Sympathy for the plight

FIGURE 4.11   Expert or strong boss may provide the vision.

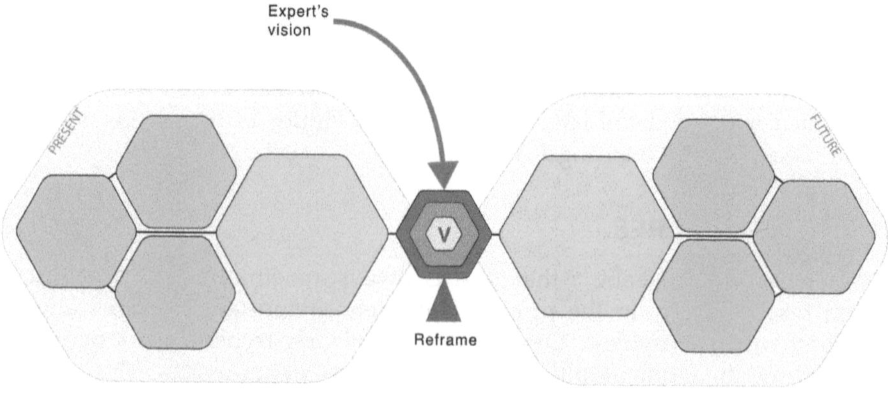

Expert's vision

PRESENT

FUTURE

V

Reframe

of underprivileged moves social entrepreneurs to extraordinary feats and achievements. Dismay at inefficiency and waste moves many conservation-alert activists to great inventiveness and achievement. The more classical entrepreneurial passion for self-fulfilment in the forms of wealth, security, status, testimonial achievement, etc. provides others with powerful drive. Where such passion, experience and drive exists within the project initiator and leader, the strength of the individual's personality and willpower often pulls the project through. Of course, the prior experience must be valid, and the leader must practise a capacity to adapt and learn, as is advocated with the Lean Startup methodology, founded by Blank. Blind, stubborn passion has no value.

CORPORATE EXPERTISE

Within established larger organisations, as we have said earlier, there is a wealth of accumulated knowledge. In effect, the practice of the ongoing business, which involves engagement with existing users and perhaps non-users, is research in its own right. Visions for improved alternatives have often come to mind and may have been discussed at length beside water coolers and over evening drinks after exhibitions and customer visits. But, as informal meanderings, they may never have been legitimised through formal description, senior executive consideration, project prioritisation or budget allocation. Sometimes, maybe by chance, one of these catches the attention of a senior manager at an appropriate time, and is raised by decree to the status of active project. By this genesis, the senior executive may present the vision as intact and not for changing. The project team has clear instruction to implement the solution to achieve the vision.

Notwithstanding senior management or founder pressures, an innovator or innovating team will serve itself well to explore around the vision to the limit of what is allowable in the circumstances. Serious errors are better exposed earlier than later, when they may more likely be wrongly attributed to poor execution rather than faulty vision! More often, the vision can remain substantially intact but materially adapted and enhanced by achieving better understanding of nuances from studying the context.

The serious problem with vision-by-decree for an innovation project is inevitability of unconscious bias. Even experts have limitations of knowledge, scope of expertise, perspective, self-understanding and hence objectivity. On the other hand, we must acknowledge that it is natural for senior executives to score highly on measures of conviction, passion and drive, which are generally nett contributors to successful project completion. So, there is, again, a fine balance to be achieved between impartiality and conviction, and it requires good leadership and good process facilitation to arrive at a really good innovation vision that relies mostly on embedded expertise. Roberto Verganti has proposed such a methodology in his recent book, *Overcrowded* (Verganti, 2016), discussed also in chapter 2. In this book he advocates a process of *criticism* or dialectic discussion, where in-house experts are encouraged to bring to the surface their own ideas for innovation visions. Through a structured process of discussion with, first, one other colleague, then with many other discussants, the multiple ideas are tested dialectically, refined, merged, compromised and enhanced until a candidate final vision emerges. Only then,

Verganti advocates that the concept for the new vision is brought to potential users for their validating input. This corresponds to the Validate stage of the ARRIVE process, discussed in chapter 6.

As we have said before and as illustrated above, innovators must take a pragmatic approach to process adherence. The 'ideal' process may not suit every context and it must be adapted to local circumstances and established operating routines. If a stage or major section of ARRIVE must be compromised, it is best to do so with full knowledge of the deficit so that this may be compensated at another stage rather than to ignore the whole process or to abandon a whole project.

In effect, a framework such as ARRIVE is a project de-risking tool. There is still the possibility of project success without adherence to the full process but, as more elements of the framework are abandoned, the risk of failure will increase.

METHODS IN THIS CHAPTER

Synthesis and reframing require reflection, collaboration and creativity. The methods in this chapter are chosen to promote these attributes and to assist in the practical task of managing and sifting through sometimes very large amounts of data to produce actionable output. We provide guidance on the following techniques in the second half of this chapter:

- Storytelling
- Affinity Mapping
- Persona
- Prioritising Insights
- Reframed Project Vision formulation
- Persona Needs Matrix

OUTPUTS AND TEMPLATES

Reframe marks a threshold from the present into the future, and its outputs are guides as to how a future reality will be created. The guiding artefacts we produce to accompany that creative activity are as follows:

- **Reframed Project Vision:** After analysing the data and synthesising patterns and themes, you know who your target user(s) will be. What is your vision of the desirable outcomes for that user in the targeted use-case context? The *Reframed Project Vision* will be your team's constant guide through the ideation, conceptualisation and validation processes to come. It represents your high-level understanding of the improved paradigm or new meaning that you intend to bring to your target users and customers. See our *Reframed Project Vision* template later in this chapter.
- **Insight Notes:** Through the course of the research and data synthesis, you encounter many 'Aha' moments and some key insights are triggered. Some of these may be real but modest in effect. Others may be more significant and after a lot of consideration with your collaborating colleagues you may realise that some of these are profound.

  While you are in full flow the insights exist in the team's collective consciousness and the dynamic assembly of notes and sketches that have evolved from the raw data. As the ideas settle, you will want to capture the key insights for later reference and communication with others. You need to capture the key actionable insights in a way that you can easily access them later. The Insight Note template helps you to record your observations, which have been the trigger for the new mental model, as well as your thoughts on why they have occurred (looking backward) and how they might give rise to specific new challenges and solutions (looking forward). See our *Insight Notes* template later in this chapter.

* **Persona Needs Matrix:** Personas represent various user types in your target context. They are trying to achieve specific outcomes and you have deduced their measures of success for reaching these aspirations. A full list of all of these needs will often reach hundreds of items, but most likely there will be a few core needs, with a small number of principal ancillary needs. It is essential that you capture these in a clear and explicit listing, which forms the requirements constraint for future solutions to be generated. See our *Persona Needs Matrix* template later in this chapter.

Reframe | methods

a audit — What is known?

r research — Find out more

r reframe — Change Perspective

i ideate — New ideas & Concepts

v validate — Test with users

e execute — Develop, deploy & scale

1

THE PROCESS OF REFRAMING
Overview of the process and sequence that will deliver the rght brief – user needs, insights, and an overarching vision.

 2

STORYTELLING
How to share individual experiences and bring to front-of-mind for the while innovation team all research findings.

 3

AFFINITY MAPPING
Make sense of large amounts of research data, by organising and sorting new data identifying the significant themes and opportunities.

 4

PERSONA
Personas provide a condensed representation of key user-descriptive research findings and a visual focus for inspiration in innovation.

 5

RECORDING INSIGHTS
Insight Notes record and elaborate on the insights that you judge have potential for giving most impact and differentiating distinction to your user.

 6

REFRAMED VISION STATEMENT
Confirm and restate the problem being addressed, and high level criteria for success. Clear reframing allows team members to be aligned in their efforts.

 7

PERSONA NEEDS MATRIX
The Persona Needs Matrix collates the main user needs and establishes a meaningful priority ranking.

In the reframe phase we synthesise the research findings into a strategic vision and an actionable brief.

THE PROCESS OF REFRAMING method 1/4

Throughout Audit and Research, your innovation team has gathered a large quantity of separate observations, disjointed pieces of information or data fragments. How do you make sense of it all? How do you synthesise relevant themes, user needs and new insights from raw research data? How do you get clarity from the haze, to focus on the innovation that brings impact?

Our objective here is to identify patterns or groups of data points that form a coherent cluster of observations, which may represent distinct user needs, aspirations, motivational drivers or behavioural patterns. This requires intuitive acts of pattern recognition and demands focused attention and collaboration, because it is difficult and mentally demanding.

Some cluster themes serve to describe particular needs and aspirations (desired outcomes), which constitute requirements for all solutions that may be developed. Needs and aspirations answer the question: *What is the user trying to achieve?*

Other themes shed light on the users' motivational drivers, behaviours and circumstances, the 'why' behind the 'what'. They provide a deeper understanding and clearer mental model of what is important for achieving the desired outcomes successfully. These are insights. Insights inspire creatively differentiated solutions that will delight and be compelling for the user. Some themes combine all of the above.

The team's immersive understanding, with newly clarified needs and insights, allow reframing of the project's overall vision in the form of a *Reframed Project Vision*.

Here, you create three defining documents for your project, which you will reference regularly as it progresses:

1. **Reframed Project Vision** – to provide clear focus
2. **Insights** – to provide leverage for differentiation
3. **Persona Needs Matrix** – to list the core jobs to be done

Step by step

1. **Before you start**
   Find a space, with free, large, 'writeable' walls, to use as a 'war room'. Gather your research collateral: journey maps, empathy maps, photographs and any other research material.
2. **Review research findings; share stories**
   Review all research findings. Share observations and research findings by storytelling, while transferring observations or micro-vignettes from the stories onto Post-its. These are data fragments, building blocks of a larger understanding.
3. **Display and start to sort the data fragments**
   Display the data fragments on a wall by speaking and placing them one at a time. Discard duplicates.

4. **Sort data fragments (Affinity Mapping)**
   This is the heart of data synthesis. It takes time and effort but it's worth it to get shared clarity on the outputs. Sort the data fragments into clusters where there is a close relationship between the various fragment topics. Provide a label to each cluster that describes its main theme.

5. **Identify insights**
   While all data fragments are relevant and useful, some clusters are particularly surprising, intriguing, inspiring, insightful, and provide a deeper understanding. We call these insights.

6. **Reframe the project brief with a new Vision**
   You know more now about the problem than when the project started. Use that knowledge to craft a new vision statement for the project.

7. **List specific needs**
   A vision provides focus. Insights provide leverage to delight. As well, every innovation context involves users trying to achieve desired outcomes. What are the jobs they are trying to do? What do they need to achieve? Use the Persona Needs Matrix to capture and prioritise detailed needs.

THE PROCESS OF REFRAMING template 2/4

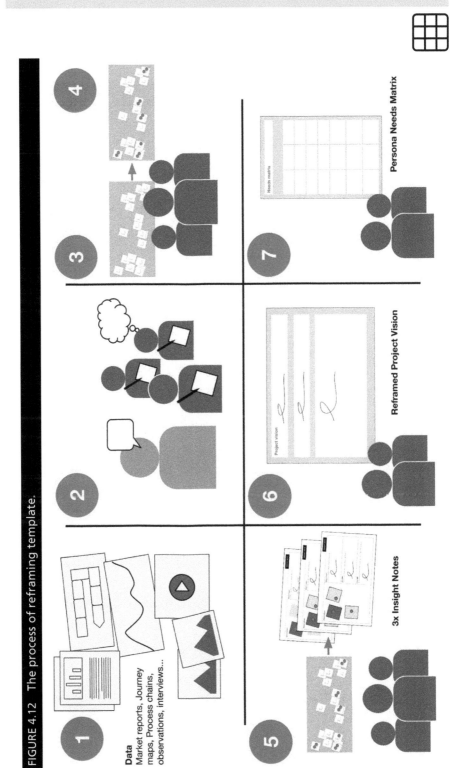

FIGURE 4.12 The process of reframing template.

FIGURE 4.13 The process of reframing example.

THE PROCESS OF REFRAMING faqs   4/4

Q What type of innovation project will benefit most from the process of reframing?

A When a problem has a high degree of uncertainty, with ambiguous data, unclear objectives and unknown root cause, it is important to choose an appropriate frame to tackle its resolution. As we have said, problem framing determines the solution space. Good reflection on the Research phase enables you to choose the frame wisely.
A common example of such a problem is where there is an underlying human behaviour and motivation at its root, and it is significantly influenced by complex and intangible emotional or social factors.

Q Are there projects where the reframing stage might be omitted?

A Reframing is not required for projects of a purely scientific nature or tending that way.
Alternatively, where a strong initial brief exists from prior research or by strong directive from the executive team, a substantial reframing may not be welcomed or acceptable The innovation team may decide to supplement the original brief with the new insights but skip the reframing piece.

Q Why choose only three insights, as in step 5? Why not more, if they are available?

A Insights are deep and meaningful. They represent a new model of understanding, which you plan to exploit in developing a meaningful total value proposition. It is hard to tackle many deep topics like this simultaneously. Also, there is a danger of cross interference, and doing all of them poorly rather than one well. It's best to work with at most three insights that are compatible, or maybe even just one. Then, later, you can revisit the others for further development, if necessary.

A process for sharing new information across the wider innovation team, this method offers the benefits of sharing raw source data and collaboratively interpreting that data.

Everyone in the team gets exposed to the richer description given by a flowing narrative, which is stimulated by the research artefacts. This is far superior to a concentrated few words or numerical metric descriptions that are prone to being inadequately contextualised and narrowly interpreted by a sole researcher. Further, team storytelling, with its discussions and clarifications, allows the combined wisdom of the team to capture a broader range of nuances and interpretations, all of which are noted in the form of 'data fragments' on Post-its.

Step by step

1. **Gather individual research material**
   Before you start, each participant must have to hand a felt-tip pen and plenty of Post-its. Also, you must display your completed journey maps, empathy maps, photographs and any other collateral to share your observations and wider research.

2. **Share stories, one researcher at a time**
   Select one researcher to start. Ask him to tell the stories surrounding his user observations and research findings. Share distinct moments of observed experience and explain the context. This process helps to immerse the team deeper into the user's reality.

3. **Capture data fragments on Post-its**
   While one member shares his stories, other team members capture interesting observations, challenges or micro-vignettes of information from the stories on Post-its. These are data fragments, building blocks of a larger, complete understanding.

4. **Repeat for each researcher**
   Repeat steps 2 and 3 with each researcher until all research has been shared, and each team member has his own stack of data fragments on Post-its.

5. **Share, display and review data fragments**
   Place all Post-its on the display wall. This is a further opportunity for collaborative stimulation.

   One at a time, team members place their Post-its on the wall, reading aloud the content as they do so. Place a new Post-it near another if it has a similar theme or meaning, but don't dwell on this for too long. All the while, members may create and place new Post-its, which might be stimulated by listening to those already placed.

When all Post-its are placed, discard clear duplicates, but be careful not to wield the axe too heavily. Do not discard a Post-it that, although similar, may have an intriguingly nuanced difference to another.

Documenting the perfect Post-it

Post-its play an important role in the synthesis and reframing process so it is important to craft the note carefully.

1. Use short phrases or sentences in the form of a 'vignette'. A vignette is a short evocative description or portrayal of an issue.
2. Use one Post-it per item of information, or fragment of data.
3. Use a felt-tip pen and write as large as possible, to fill the space. It must be visible to the whole team standing up to 2 metres away.
4. Consider distinguishing different types of information by colours of Post-it. For some people, and in some circumstances, this may cause a cognitive overload. If so, don't push it! Other circumstances may allow it. For example, Jon Kolko, the noted Director of Austin Center for Design, recommends you distinguish with colour between two types of data: market patterns and market insights.

Market Insight

A data point or piece of information that suggests a new understanding of someone or something.

Market pattern

Repeated occurrence (trends) of market behaviours.

Statistical facts are great here!

FIGURE 4.14 Storytelling process template.

STORYTELLING   example   3/4

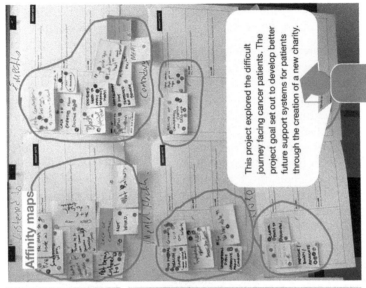

Affinity maps

This project explored the difficult journey facing cancer patients. The project goal set out to develop better future support systems for patients through the creation of a new charity.

Journey map

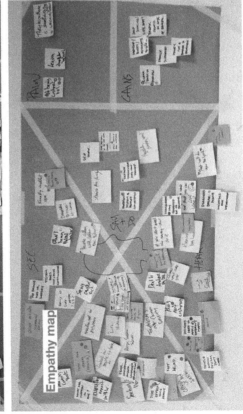

Empathy map

FIGURE 4.15   Storytelling process example.

Q Should I use props while telling the stories of my research?

A Yes, wherever possible. You should have your array of research collateral beside you: journey maps, empathy maps, photos, infographics, notes and other artefacts.
By regularly referring to these, you will more easily remember interesting details and it will help the listener better to understand the context. Anything that helps the story to resonate with the listeners is worth including.

Q How much detail should I include when sharing stories?

A Don't be afraid to discount portions of information if you don't think it is important to the project. Remember there will be many stories by the time the whole group has had a chance to share all research.
Each individual will make judgement calls on what to include so as to keep the process manageable. Prioritise the surprising insights and compelling problem areas, as these provide the key opportunities for innovation.

Q It must take a long time for all 'data fragments' to be shared with all teammates by storytelling. How long would you expect to spend at this?

A Of course, you and your team have gathered a lot of data. Discussing it all in detail would take a very long time! But, that's not what is intended.
The purpose of the 'storytelling' stage is to share, not discuss! Minor discussions to clarify a point are allowed, but there must be no attempts at persuasion or argumentation. In addition, a member may interrupt the flow if he has a data fragment similar in meaning or topic to one just posted, so that he can place the second Post-it physically near to the first.
Everyone gets a chance to express their preferences later, through voting.
Keep moving smoothly, at a constant pace through the Post-its, without side discussion or digression, and you'll find the data sharing will be completed between one and four hours for most sizes of project.

This is a method to manage large amounts of qualitative data (observations, statistics, opinions, emotions, other issues) and make sense of them by recognising patterns and organising into groupings, or clusters.

In a cluster, all constituent data fragments have a close relationship based on a meaningful pattern of features, behaviours, actions, objects of action, or other criteria. The pattern is emergent. That is, it is unknown from the start and arises creatively in the minds of the researchers.

In an innovation project, the emergent patterns or themes that you derive from research are potentially valuable discoveries that can provide inspiration and leverage for doing something special or unexpected for the target user. Usually, the clusters are many and varied in topic and emphasis. Hence, the first task is to select which ones will deliver maximum impact and differentiating delight to the user. These are insights.

The roots of Affinity Mapping

Affinity Mapping follows principles of open coding often used in qualitative social science research. Coding is a means of organising data and interpreting its meaning. In particular, open coding is a form of grounded theory, a process of theory building that is grounded in observation of repeating patterns.

Open coding is conducted without predefined themes and structure. The meaning and relationship of data are evolved through continuously asking questions such as:

- What is the underlying issue and phenomenon?
- Who are the actors and what roles do they play?
- How does the phenomenon occur?
- When, for how long and where does it occur?
- Why does the phenomenon occur? For what intention or purpose? Is it a consequence of overarching goals, strategies or tactics?

STEP BY STEP

1. **Team size and duration**
   Affinity Mapping requires a state of flow in the team, free from distractions. Free interaction between team members assists positive reinforcing of productive ideas and filtering out of inferior ones. Team sizes of 3–6 work best for lively discussion and diversity of pattern recognition. Expect a duration of 1–3 hours for most projects.
2. **Physically move the Post-its into clusters**
   After storytelling, your data display can appear a daunting, chaotic mess. Examine the data fragments, methodically and iteratively reflecting on their meaning and implications, and seek out patterns or connections between them.

Take one Post-it, or existing mini-cluster. Look for other Post-its that relate to it and move them adjacent to it. When that cluster is exhausted, take another (unassigned) Post-it, and try to build a cluster around it. As the clusters develop, give each one a name to represent its theme. At any time, don't hesitate to re-assign a Post-it, merge multiple clusters, divide a cluster into multiple ones, or make other changes as appropriate. If a data fragment belongs in two clusters, duplicate it.

3. **Keep discussions compact and useful**

This is an organic process. It is common to have differing opinions among team members about the cluster relationships and where a Post-it should be placed. The ensuing discussions are very positive and generative, and they help develop deeper understanding. However, don't let disagreements dominate the process. The team leader must be prepared to impose decisions after a reasonable time, if necessary, and to drive the process forward.

4. **Agree cluster names; keep orphans aside**

Eventually cluster formation saturates, with no more identifiable themes. You will certainly have isolated Post-its remaining, which you should leave to one side. We call them 'orphans'. Agree the final name for each cluster.

AFFINITY MAPPING template 2/4

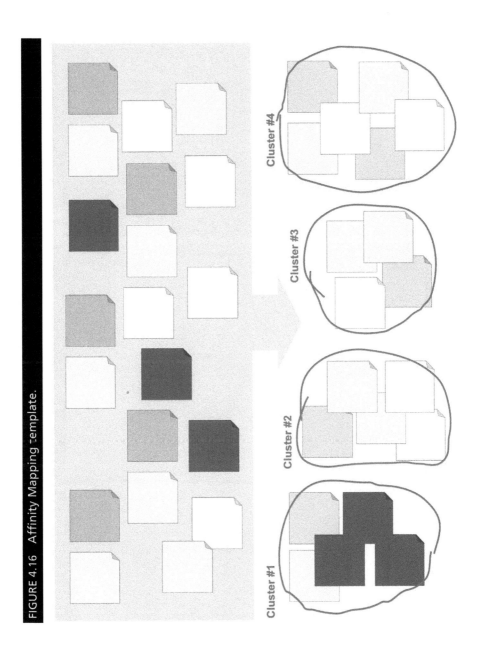

FIGURE 4.16 Affinity Mapping template.

Cluster #1

Cluster #2

Cluster #3

Cluster #4

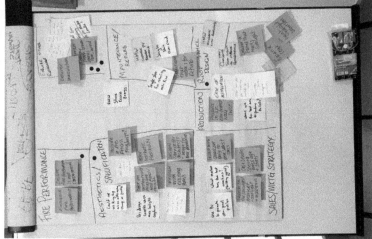

FIGURE 4.17   Affinity Mapping example.

Q How many clusters should we go for?

A There is not a universal answer. A narrowly focused project will have fewer clusters, as will a more substantial project where the research has been done only superficially. Better research gives rise to richer data and this may result in more cluster themes to be considered, with themes being richer and more convincing.
Having said that, fewer than three clusters is rare. It's hard to imagine a project that has only one or two dimensions. On the other hand, more than 12 clusters, for example, even for a large project, appears to be parsing too finely at this stage and you should consider broader interpretation of a reduced number.

Q Are there standard cluster themes that we should always look for?

A No, themes are specific to the context and emerge from the data. You should expect to find them spanning functional, social and emotional aspects of the context under study.

Q What should we do if:

(a) two Post-it notes are very similar, or

(b) a Post-it note belongs in two clusters?

A (a) If two Post-it notes are very similar in meaning, consider discarding one, or combining both observations into a new note.

However, be careful not to lose important information, or nuanced understanding when you discard one or the other.

(b) If a Post-it message contributes to more than one cluster, simply replicate it and place it in all relevant clusters.

Be careful: if a message appears to be relevant to more than two clusters, then it may be that the cluster themes are not adequately distinct from each other. Review the cluster themes to resolve this.
Alternatively, the Post-it message may be too generic and therefore providing only noise to the process. Consider leaving it out, instead.

This is a fictional character built from user research to create a vivid representation of a customer.

A persona is a fictional character, created as a synthesis of your research, that is used to represent a potential user of your product, service or process. You may require one or many personas to represent your users, depending on the complexity and variety of your target context.

The concept of persona in design was originally introduced by Alan Cooper. Cooper (2004) describes a persona as a precise description of a potential user and what he wishes to accomplish. Personas are exemplars of specific groups of real consumers who share common characteristics and goals. They represent an aggregate of target users who share common behavioural characteristics (Pruitt and Adlin, 2006). Note, especially, that a persona is a precise and specific exemplar; it is not an average catch-all representation.

Without personas, design teams can easily stray away from focusing on the target consumer and may end up drawing instead merely from their own experience and assumptions. Personas may be as simple or complex as you feel appropriate, and may include details such as psychographic profiling, brand preferences, behavioural patterns, social and professional activities.

For many projects, we recommend starting with a simpler 'lean' persona as outlined here. As your knowledge and experience develops, move on to more in-depth persona constructions.

Remember, a persona is an evolving construction. You should validate and update it as your focus and your understanding evolve.

Step by step

1. **Collect and analyse data**
   Creating a persona begins with collecting and analysing good research data. This data consists of interviews with users as well as data from your experience and empathy maps.
2. **Form a hypothesis**
   As a team, review the research to find patterns in the data that suggest grouping similar people together into types of users. Agree on one or a number of user types you plan to focus on, but not too many.
3. **Add detail**
   Add rich detail. It is important to make the persona realistic, believable and relatable; so, give him a name, title, sex and age. It is valuable also to represent the persona visually with a photograph or sketch. Find an image that shows context and conveys emotion; it must appear as a real person.

Next, fill out all sections on the template overleaf. Remember, this is not a product of your imagination. It is drawn wholly from your user research. Once you have completed the life and scenario sections you can move to the Goals, Frustrations, Motivations and Quotes.

Goals: What goals is the person trying to achieve? These may be specific to the context of your proposed product as well as broader life goals. They may be tangible (a new service or a mechanical aid) or intangible (better productivity, or reassurance).

Frustrations: What is preventing the person from achieving the goals?

Motivations: What motivates your persona to take action, to do what they do and to want to do it better? How do they make decisions? What do they fear?

Quotes: Quotes are a powerful way of helping the team relate to the persona. Draw quotes from your research and, if necessary, mix them up for impact.

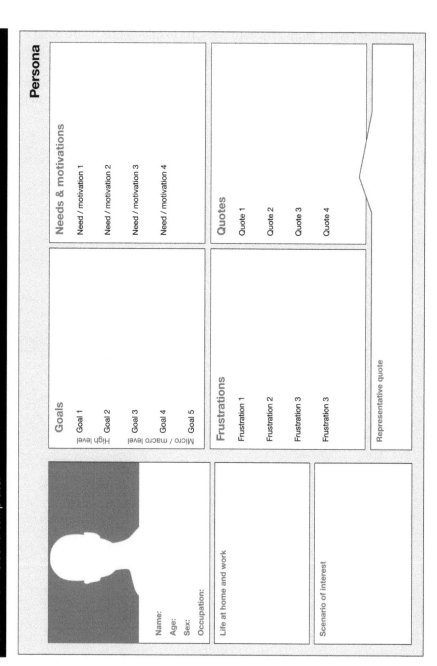

FIGURE 4.18 Persona template.

Persona

Needs & motivations

Need / motivation 1

Need / motivation 2

Need / motivation 3

Need / motivation 4

Goals

Goal 1

Goal 2

Goal 3

Goal 4

Goal 5

High level

Micro / macro level

Frustrations

Frustration 1

Frustration 2

Frustration 3

Frustration 3

Quotes

Quote 1

Quote 2

Quote 3

Quote 4

Representative quote

Name:

Age:

Sex:

Occupation:

Life at home and work

Scenario of interest

PERSONA example

FIGURE 4.19 Persona example.

Persona

Age:

Sex: Female

Occupation: Product Manager

Life at home and work

Josephine is a 34 year old busy product manager in a Dublin city center tech company. She is married with one young child. Her husband is a software developer in another city-centre business...

Scenario of interest

Josephine has flexibility in her job, and for two days a week she can work from home. She loves this flexibility, but often finds she is not very productive...

Goals

High level

Goal 1	To balance work and family life, while excelling in both.
Goal 2	To one day start her own business in tech with an idea she has been thinking about.

Micro / macro level

Goal 3	To have a 3 hour "clear run" at tasks that require headspace
Goal 4	To feel satisfied that she is working to a high level and not just getting things done.
Goal 5	

Frustrations

Frustration 1	Constant distractions at work
Frustration 2	Permission to work from home but not able to make it happen regularly
Frustration 3	Not being able to do work to the best of her abilities
Frustration 3	Her manager is not organised and often gives her unrealistic deadlines

Needs & motivations

Need / motivation 1	Spend time with her young child and be a great mother
Need / motivation 2	Josephine loves having interesting projects to work on
Need / motivation 3	Remove or limit distractions in her work
Need / motivation 4	Have time to think and do deep work
	Look after her physical and mental wellbeing

Quotes

Quote 1	"You know, it's really hard to just get stuck in!"
Quote 2	" Sure, what's another deadline!"
Quote 3	"I wish people would just give me 10 minutes to get this done right"
Quote 4	

Representative quote

"Sorry, Babe, looks like it will be another late one tonight... I promise I will be home early tomorrow night."

Q Why do I have to create a persona? Is it not better to have a list of user requirements?

A A list of user requirements or technical information has no integrating narrative, and it is too easy for the innovation team members to revert to their own assumptions about the customer. Lists lose sight of the end user and all the personality that lies within, which provides the contextual richness.
It is too easy to forget that we are innovating for a living, breathing human being. Personas are also a great way to ensure the entire team is focused on the same 'person', not pulling the project in different directions.

Q How many personas should I create for my project?

A How many different types of users are there? Some contexts have one clear user type and others have many.
Remember, while a persona is constructed as if it were a real, single person, in fact it is representative of a type (of user). It is not helpful to try to distinguish between different types with too fine granularity. A persona's purpose is simply to inspire.

It is not uncommon to have two or three personas. More than six is probably too many, unless it is a particularly complex project.

Q The persona I created feels made up. What am I doing wrong?

A It is a common mistake to believe you can simply make up the persona details from your own imagination, even if you are expert in the field. Your own personality and biases will inevitably intervene, dominate, and corrupt.
A good persona is created from rigorous and rich user research. Your persona should be solidly, clearly and traceably connected to your research data.
Towards conclusion of the persona building, have your team engage formally with some customers, to validate and make sure the persona you have created is realistic and representative.

RECORDING INSIGHTS method   1/4

When Affinity Mapping has sorted your data into thematic clusters, instead of dealing with data fragments you have multiple strands of knowledge represented on your display. Though they vary in scope and importance, all are relevant to your complete understanding. However, some are more powerful than others. A few of them, you may judge, lead to a whole new model of understanding that may have potential to leverage a high impact and delightful experience for the user. Now, your team uses its best judgement to choose these high-impact insights.

One high-quality insight provides more potential than a number of less remarkable findings. High-quality insights usually spawn a combination or a series of innovation features, so you should seek only to develop a small number, or even one. Typically, initially targeting three good quality insights is plenty.

An insight is a shift to a new and better understanding, which may provide a route to innovation or a problem's solution that goes beyond the ordinary or superficial. The channel for the insight recognition is an innovator's creative mind, honed by intuition and expertise, so it is more productive and effective to identify and prioritise insights using the team's combined intuition and expertise, rather than by a team-leader's diktat. Of course, in practice, there may be distinctions of expertise or corporate seniority amongst people in the room. Completing the Insight Note described here helps tease out stated preferences and reduce arbitrary influence and bias.

Step by step

1. **Review and discuss all clusters**
   Take each cluster from your affinity map and have the team reflect on what it is saying. Use the Insight Note headings as a guide for the discussion. In particular, you want to determine if the cluster represents a new perspective of understanding of the research area, or a section of it. Additionally, does it give rise to an opportunity to add real impact to the user? Each team member keeps his own notes.
2. **Shortlist the clusters**
   You may have as few as two or three clusters, or you may have more than ten. If more than three, we recommend shortlisting to the three most likely to provide greatest impact and delight. Let the team do this by power-dot voting. Choose the three most popular clusters for more detailed analysis.
3. **Build your understanding of the insights**
   Addressing each of the selected clusters one at a time, transfer the Post-its of the cluster to the left-hand side of an Insight Note. Encourage discussion and reflection by the team focused on completing the right-hand side sections of the note, in turn:

   • **Why?** Why do these observations stand out as significant or curious? Do they show an anomaly, impasse, an unexpected coincidence or pattern? Do they align with a macro trend?

- **What?** What is the new understanding or mental model you now have? Keep your description as simple and factual as possible. Don't elaborate beyond what is necessary to describe it clearly.
- **How?** How might this insight be used to create a great solution for the user? Does it give rise to clear challenge(s) that can be expressed in the *How Might We* form?

The completed Insight Notes are your considered and elaborated insights, for use as levers to delight your user with a great and unexpected solution.

4. **Keep all data for further processing**

   The remaining clusters and individual Post-its contain further information about user issues that need resolution. You and your team have decided that they may not have the same power to surprise and delight as your chosen insights. Nevertheless, they represent, in various ways, real needs where good solutions are desirable, expected, possibly essential. These needs will be captured by the Persona Needs Matrix, which we describe later.

RECORDING INSIGHTS template 2/4

FIGURE 4.20 Insight Note template.

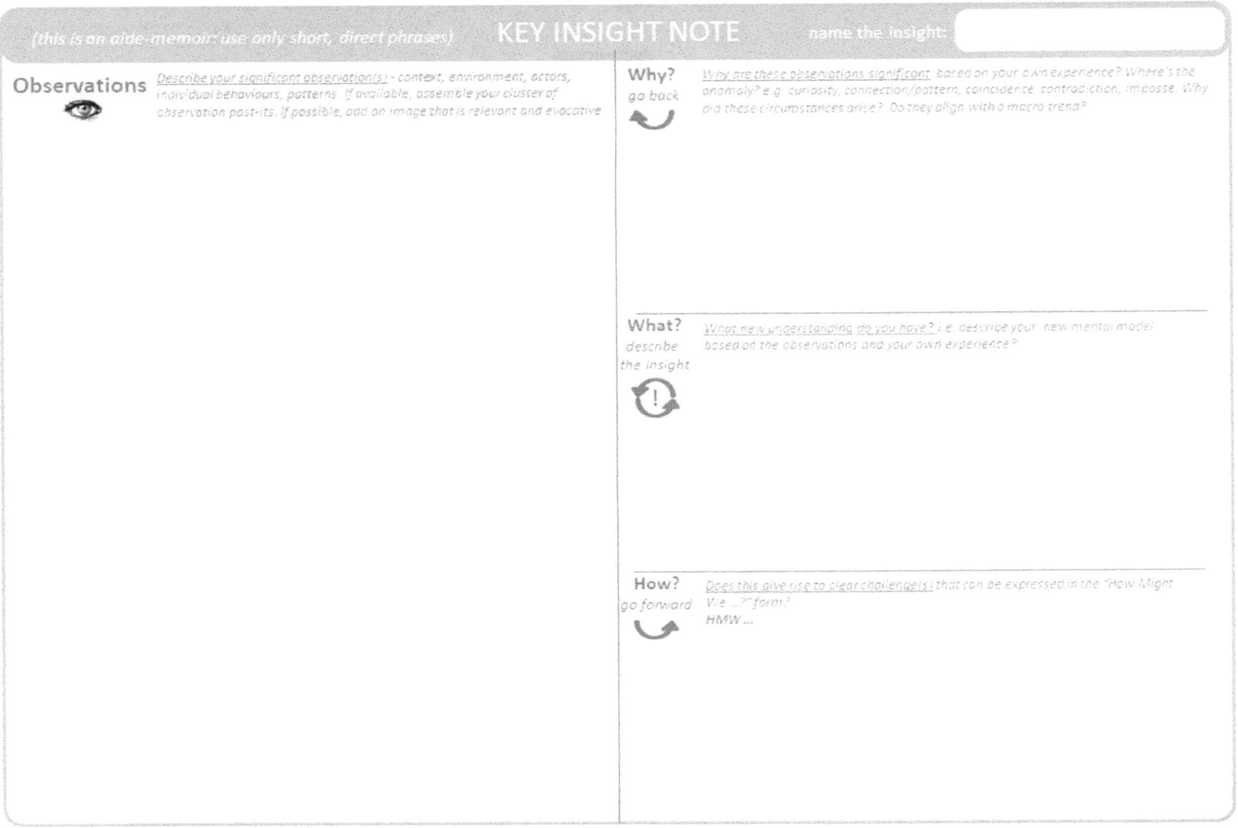
(this is an aide-memoir: use only short, direct phrases)
KEY INSIGHT NOTE
name the insight:
Observations
Describe your significant observation(s) - context, environment, actors, individual behaviours, patterns. If available, assemble your cluster of observation post-its. If possible, add an image that is relevant and evocative
Why?
go back
Why are these observations significant, based on your own experience? Where's the anomaly? e.g. curiosity, connection/pattern, coincidence, contradiction, impasse. Why did these circumstances arise? Do they align with a macro trend?
What?
describe
the insight
What new understanding do you have? i.e. describe your new mental model based on the observations and your own experience?
How?
go forward
Does this give rise to clear challenge(s) that can be expressed in the "How Might We ...?" form?
HMW ...

FIGURE 4.21 Insight Note example.

KEY INSIGHT NOTE — name the insight: Socialising adult gaming

Observations *[this is an aide-memoir: use only short, direct phrases]*

Describe your significant observation(s) - context, environment, actors, individual behaviours, patterns. If available, assemble your cluster of observation post-its. If possible, add an image that is relevant and evocative

- People are more health conscious and want to add activity to their lives

- "He was spending so much time gaming that we had no time together."

- People stop playing games as they transition to family life

Why? *go back*

Why are these observations significant based on your own experience? Where's the anomaly? e.g. curiosity, connection/pattern, coincidence, contradiction, impasse. Why did these circumstances arise? Do they align with a macro trend?

Sole player games cause a conflict of guilt v. pleasure for the player. Traditionally there has not been a good trade-off.

There is a large fall off in the purchase of game consoles when people reach 30 years.

What? *describe the insight*

What new understanding do you have? i.e. describe your new mental model based on the observations and your own experience?

People enjoy interactive, social games and often feel guilty after spending time playing alone.

People feel good about themselves after physical activity.

How? *go forward*

Does this give rise to clear challenge(s) that can be expressed in the "How Might We ...?" form? HMW —

HMW create more whole-family entertainment system

HMW create an entertainment system in e-gaming tradition that combines with physical activity

This example draws on the amazing insight behind the Nintendo Wii. As Nintendo failed to capture market share against Playstation and Xbox the team resorted to considering adjacent market segments for new opportunities. The Nintendo Wii went on to be the most successful selling console for many years.

RECORDING INSIGHTS   faqs   4/4

Q Why bother completing an Insight Note? Is it not sufficient to record or take a photo of the cluster of Post-its?

A Remember, insights are precious! You have chosen up to three that will be the foundation of your new approach to the project from now on (Reframe). It is important that you cultivate a full understanding that is shared by the team.

When the team discusses the insight, the understanding deepens and is shared by all. Ambiguity and fuzziness are minimised by defining clearly and simply what is the new understanding and its origins. The major ensuing project challenges are captured in the HMW questions.

In addition, there's another very practical benefit. Even if the understanding of the insight is fully clarified, it probably will diffuse in memory with time as other project factors crowd into the team's consideration.

It is comforting and valuable to have the clear understanding of a cornerstone of the project's advancement clearly captured for future reference.

Q Are there any rules for creating How Might We questions from insights?

A One insight may give rise to multiple challenges, and each must be captured in HMW format. The HMW statement must:

i. specify the desired outcome to the challenge from the user's perspective, not the means (solution) to achieve it.

ii. be broad enough that it allows for a variety of possible solutions, yet focused enough to facilitate actionable ideation of a solution.

Q Why do you recommend a maximum of three insights?

A One high-quality insight might spawn a whole new innovation direction, with distinctive challenges and set of new features. One good insight can provide more potential than a number of less remarkable findings. Three such insights are as much as any team can handle, because team performance suffers when its attention is spread too thinly.

Every well-executed project is led by a clear, guiding vision. While the project may have started with an uncertain, hazy or abstract goal, the Reframed Project Vision is the revised view of the project's high-level desired outcomes.

By now, you have researched extensively and have learned a lot through direct observation and reflection. You have analysed and synthesised your research data to get greater understanding and deeper empathy for the user(s), their needs and desired outcomes. In previous methods in this chapter, you have constructed detailed persona profiles as representatives of your main user types and you have selected up to three key insights where you judge you may achieve the greatest impact to benefit them. By doing so you have, in effect, selected a new frame through which you will look for great solutions. Capture this frame in a Reframed Project Vision template.

The Reframed Project Vision provides a guiding 'North Star' to the innovation team throughout subsequent stages of development. By clarifying who it is that the innovator intends to serve and the overarching desired outcomes in the specific context, it ensures team members stay aligned and focused. Hence, the Reframed Project Vision is centrally important for the project and merits thorough, unhurried consideration.

Step by step

1. **Review your data synthesis results**
   Spend at least one hour reviewing all data clusters and isolated data fragments and, in particular, the personas and three evolved Insight Notes. Meanwhile, the team leader or other designated 'interpreter' captures his best interpretation of a Reframed Vision Statement representing the orientation of the team's discussion.

2. **Share the draft Reframed Project Vision**
   Transcribe the draft Reframed Vision onto a whiteboard. Remind the team that this is an important exercise in strategic judgement, which will impact the remainder the project. Do we have a Reframed Project Vision that, if realised, is likely to provide real and meaningful change to the user's experience and that is deliverable by your team and the firm?

3. **Step through all sections of the template in turn**
   'Our user(s)': Clearly identify the user(s). The personas were developed for this purpose. Append these persona boards to the vision statement. Alternatively, with a dominant single user type, some vision statements directly describe the user. Use empathic language. Think how you would describe a friend, for whom you are about to create a gift of improved experience. Use colloquial phrasing that is more than a stark, lifeless demographic classification or functional fact. Evoke a sense of the user's personality or culture. Strengthen the emotional identification by applying a picture of the user.

   'in this context': Set the scene by describing the circumstance, system, process or paradigm that is problematic. When or where does the user find a difficulty, impasse, or need for improved experience?

'need(s) to achieve the key outcomes …': State the users' overarching goals or motivations, as you have understood them following research. Assign priority to them, based on your intuitive understanding of their needs as well as your ability to impact them.

- What are they trying to achieve? Use verbs.
- What difficulty, impasse or poor experience requires improvement?
- What are the functional/social/emotional jobs-to-be-done, at a high level?

Describe desired outcomes, not solution methods! Be selective. Don't overload. Avoid the basic, banal and obvious. Give special attention to the motivational insights that might help leverage a distinctive offering.

4. **Conclude with a definite vision**
   The Reframed Project Vision is likely to evolve in content and tone throughout the session, in response to team contributions. Conclude the session when there is no further substantial evolution, and team agreement is clear. In cases of stalemate, a decision-call by the senior member may be necessary.

REFRAMED PROJECT VISION template 2/4

FIGURE 4.22 Reframed Project Vision template.

REFRAMED PROJECT VISION

project

Our user(s)

Personas OR
Describe the person(s).
Use empathic language
& refer to personality,
lifestyle, professional
and social status.

Persona 1	Persona 2	Persona 3
photo	photo	photo
name:	name:	name:
representative quote:	representative quote:	representative quote:

in this context

Describe the circumstances, system, process or paradigm where the user(s) experience(s) difficulty, an impasse, a need for improved experience.

need(s) to achieve these key outcomes:

They need
to be able to ...

Describe what they are really trying to achieve, at your deepest level of insight and understanding. Distil all detail to leave only the most important. Use verbs to describe the achieving of outcomes. Use combined or separate outcomes for the various personas, as appropriate.
What are the underlying, key functional/social/emotional jobs-to-be-done?
What difficulty, impasse or poor experience requires improvement?

REFRAMED PROJECT VISION example 3/4

FIGURE 4.23 Reframed Project Vision example.

REFRAMED PROJECT VISION	project

Our user(s)
Personas OR
Describe the person(s).
Use empathic language
& refer to personality,
lifestyle, professional
and social status.

Joe is a bright and sociable software developer, in a well paid job with a multi-national company in downtown Dublin. He is aged 31 and he has recently moved into a new apartment with his wife, Jessie, and 18-month old baby, Jean. He adores his wife and young baby and is proud of what he has achieved in his life so far. He enjoys playing video games as a form of mental relaxation and turning off from work.

in this context

Describe the circumstances, system, process or paradigm where the user(s) experience(s) difficulty, an impasse, a need for improved experience.

When Joe is not at work, he likes to turn off and enjoy life with his family. There are so many things he'd like to do and often it seems there just isn't enough time.

need(s) to achieve these key outcomes:

They need to be able to ...

Describe what they are really trying to achieve, at your deepest level of insight and understanding. Distil all detail to leave only the most important. Use verbs to describe the achieving of outcomes. Use combined or separate outcomes for the various personas, as appropriate.
What are the underlying, key functional/social/emotional jobs-to-be-done?
What difficulty, impasse or poor experience requires improvement?

- Spend quality time with Jessie and Jean, which he really values and enjoys.

- Relax with his video games and still have enough time for other social mixing.

- Have a clear conscience playing games and not worry about leaving Jessie alone with the baby or his parents when they come to visit.

- Keep active, because he knows that his life at work and his favourite pastime are sedentary.

- Find a way to organise his time to do <u>everything</u>, because you only live once.

Q Is the Reframed Project Vision just a quick summarising of where we are at?

A In a way, yes. But, it is much more than that. It is a strategic reflection and positioning, that gives direction to the rest of the project.
For that reason, we recommend that you dedicate a team session to it alone. Ensure your team takes a good break, or comes afresh the next day, to your project 'war room' with all your data on display.

Q How much detail should I include in each section?

A We recommend one sentence for each key idea. Keep it tight and directly to the point. Do not mistake the high-level nature of the vision for permission to make generic, light-meaning statements. Each line must be carefully crafted to capture the essential point, going beyond the obvious where possible.
Be insightful, even surprising, where you have something new or a new way of interpreting existing information.
Finally, it should always represent the user's point of view.

Q Is it okay if the vision statement repeats elements of the insights and How Might We (HMW) questions that we have developed earlier?

A Yes. The insights and associated HMW questions lead inevitably, almost by definition, to the new vision for a solution to the users' dilemma. So, some of the concepts and phrases will be repeated. But, perhaps not all.
The insights and HMW questions are more specific, granular and varied. The Reframed Project Vision is high level, conceptual, overarching and integrating.
The HMW statements provide the immediate focus when carrying out ideation exercises, while the vision statement sets the context and should be referenced periodically to ensure staying on track.

PERSONA NEEDS MATRIX method   1/4

A Persona Needs Matrix captures and prioritises all user needs to be considered in the solution. After Affinity Mapping, your project wall is full with research data in the form of Post-its, mostly arranged into clusters of coherent themes, while some remain as isolated data fragments. All of this data provides information on what the user is trying to achieve. You have prioritised up to three of the themes as *key insights* with potential to leverage a really distinctive value proposition. Now, the Persona Needs Matrix provides a summary dashboard of findings in respect of the broader, full range of user needs and desired outcomes.

In keeping with our general approach, we focus here on the desired outcomes viewed through the lenses of the personas developed earlier in this chapter. These are the primary driving needs of the project. Of course, there are other need types as well, such as regulatory requirements, technological specifications, best and standard practice. Though basic and important, they are secondary to the persona needs and we do not treat them explicitly here.

Step by step

1. **Identify and characterise all persona needs**
   This initial stage may take some hours to complete and is best done by a small group, e.g. team leader and one or two other members, in order to keep the process manageable. Need descriptions may be conveniently recorded in a computer spreadsheet, for sharing with the team later.

   Review the 'wall' of synthesised data that remains after the key insights have been extracted. Selecting one cluster at a time, note its dominant theme and identify all distinct needs incorporated in the cluster data. Repeat for all clusters and the three key insights, as well.

   Complete a need description for each distinct need. Note the main theme with which it is associated (there may be more than one); specify the need in one phrase or short sentence; identify how you will know if you have succeeded in satisfying the need, i.e. the user's success metric.

2. **Apply priority scores with persona perspectives**
   Reconvene the full team and consider each need, from the perspective of each target persona in turn.

   The best way to achieve this is to have every team member adopt the mindset of just one persona for the whole exercise. Thus, an individual will not have to switch personas, which would be confusing. All personas are distributed among the team.

   For each need in turn, decide a single score for each persona, which represents the importance of that need being satisfied when creating a value proposition that is compelling for that persona.

Use the following scale:

3	=	essential
2	=	very desirable; would be missed if not there
1	=	nice-to-have; might not be missed
0	=	indifferent; user won't care if it's there or not
-2	=	useless clutter; negative distraction

3. Calculate the priority ranking

Sum the scores and priority rank the needs. You might choose to translate these needs explicitly into HMW questions to do some direct brainstorming, or they might be integrated during concept building (chapter 5).

4. Get more validation from users and complete it

As usual, you should do a sanity check with some real target users. Invite a sample set of users to participate and direct them to number the needs in order of priority. If the list contains more than 10, as is likely, you may wish to run a card sorting exercise. See FAQs, page 4 of this method section.

Complete the Persona Needs Matrix by confirming a final priority ranking of needs, based on informed and intuitive discussion, or by mathematically averaging the rankings from the various sources.

PERSONA NEEDS MATRIX template   2/4

FIGURE 4.24   Persona Needs Matrix (a) full template (b) need-strip.

Persona Needs Matrix									
Summarise outcomes your personas are trying to achieve. Add needs until all are identified.			project name					date	
					scores				
Theme	#	Need	Success metric	Persona 1	Persona 2	Persona 3		Sum of scores	Rank
What is the high-level need category?	Need id	What 'desired outcome' is the user trying to achieve?	How will the user measure success?						

Scores legend (vertical): 3 = essential; 2 = very desirable; 1 = nice-to-have; 0 = indifferent; -1 = useless clutter; -2

Persona Needs Strip		
Summarise a user need, one per strip. What outcome is the user trying to achiev?	project name	date
Theme	**Need**	**Success metric**
What is the high-level need category?	What 'desired outcome' is the user trying to achieve?	How will the user measure success?

PERSONA NEEDS MATRIX   example   3/4

FIGURE 4.25   Persona Needs Matrix for a showering enclosure study.

Persona Needs Matrix
Summarise outcomes your personas are trying to achieve. Add needs until all are identified.

Shower enclosure study   June 2019

Theme *What is the high-level need category?*	# *Need id*	Need *What 'desired outcome' is the user trying to achieve?*	Success metric *How will the user measure success?*	Persona 1	Persona 2	Persona 3	Sum of scores	Rank
Physical shower environment	1	Comfortable, spacious in-shower space	Floor area; Shower height; Shower head height	3	3	1	7	2
	2	Comfortable, safe access	Step height; Door width	3	1	0	4	10
	3	Ergonomically & aesthetically available showering accessories	Tidiness and convenience: Soap bar, shampoo bottles x 3, loofah; nothing on the floor	3	2	0	5	8
User wellbeing	4	Feeling refreshed	Good mood, ready frp the day!	3	3	3	9	1
	5	Relaxation and de-stressing	No in-shower stress factors: e.g. water pressure, water temperature, steam build-up. Pleasant and positive aesthetics	2	2	-2	2	11
Maintenance	6	easy weekly cleaning	5-min clean; low levels of scale build up, drain blockage, dirt trap corners	3	2	2	7	2
	7	no necessity for daily cleaning	simply walk away, and it appears clean	2	2	2	6	4
Purchase and installation	8	Style and colours to match bathroom design	Wide range of contemporary and classical colours and textures	2	3	0	5	8
	9	Short delivery time	Days to installation (max 5)	1	2	3	6	4
Installation	10	Quick one-step installation	Under 3 hours; no dirt; fully sealed	1	2	3	6	4
	11	DIY installation is easy	Clear no-confusion instructions. Back up help available.	2	2	2	6	4

-2 = useless clutter; 0 = indifferent; 1 = nice-to-have; 2 = very desirable; 3 = essential

This project set out to reinvent the showering experience. There were a number of unspoken needs unearthed, in particular emotional needs, which gave rise to a range of new feature possibilities that needed careful curation and

Q Why should we include the insights in the Persona Needs list? Is this not repetition?

A Or, you could say that you can't have too much of a good thing! As we have often repeated, insights are special and precious. They operate at two levels:
On the one hand, you have identified them as representing a new model of understanding of the user and the context of study. This holistic understanding serves as inspiration for creating a great solution to delight the user.
On the other hand, there are some clearly identifiable disparate needs embodied in the insight, and you want to include these here in the Persona Needs Matrix also, at least as a matter of housekeeping, bringing all the needs together in one place.

Q How does card sorting work?

A 1. Create a separate 'need-strip' for each need, as a cut-out from the Persona Needs Matrix, one need-strip per row.
2. Ask each test user to arrange the set of need-strips in order of importance.
3. Take the opportunity to ask why the user chose these priorities, and deepen your understanding of their needs.

4. Give them an opportunity to add any missing need-strips.

Now you have a needs ranking to compare with other user rankings and with your team's persona-based rankings.

Q It feels like we could go on forever. At what point should we stop listing user needs?

A There is a wealth of expert and common knowledge that is brought to every project. You cannot list it all, and you should not try to do so. You are entitled to assume that certain common practice, best practice, regulatory and legal requirements will be incorporated by default.
The purpose of the Persona Needs Matrix is to supplement the previously identified insights in order to steer the project through the desired outcomes and constraints of the particular case under study. While insights were chosen as having special possibility for delighting the user, the Persona Needs are more basic though usually no less essential.
Keep listing needs until you can think of nothing else that is special to this project. You may have between 10 and 50 such needs, depending on the project size.

BIBLIOGRAPHY

Argyris, Chris (1983). Action Science and Intervention. *Journal of Applied Behavioral Sciences*, Vol. 19, No. 2, pp. 115–140.

Argyris, Chris (2004). *Reasons and Rationalizations – the Limits to Organizational Knowledge*. Oxford University Press.

Argyris, Chris and Schön, Donald (1978). *Organisational Learning: A Theory of Action Perspective*. Addison Wesley.

Christensen, Karen (2008). Thought Leader Interview: Chris Argyris. *Rotman Magazine*, Winter edition.

Cooper, Alan (2004). *The Inmates Are Running the Asylum*. Sams.

Csikszentmihaly, Mihaly (1990). *Flow: The Psychology of Optimal Experience*. Harper & Row.

Dorst, Kees (2015). *Frame Innovation*. MIT Press.

Gray, Dave (2016). *Liminal Thinking*. Two Waves Books.

Klein, Gary (2013). *Seeing What Others Don't: The Remarkable Ways We Gain Insights*. Nicholas Brealey.

Kolko, Jon (2011). *Exposing the Magic of Design: A Practitioner's Guide to the Methods and Theory of Synthesis*. Oxford University Press.

Pruitt, John and Adlin, Tamara (2006). *The Persona Lifecycle*. Morgan Kauffman.

Verganti, Roberto (2016). *Overcrowded*. MIT Press.

Ideate

"Chance favours the prepared mind."

Louis Pasteur (1822–1895)

The Reframed Project Vision is the North Star and each insight provides inspiration for fresh and compelling ways to bring something special to the innovation opportunity. The Persona Needs Matrix provides the minimum set of user conditions to be satisfied by the concepts you are about to create, and consequently they will form the supporting framework for concept screening and selection.

While focusing on the direction provided by the outputs of Reframe, use a variety of creativity-stimulating techniques to ideate multiple new ideas. Each idea, at first, is a distinct stand-alone fragment of a solution. Subsequently, these fragments are manipulated and grouped to form integrated, coherent concepts, which you will bring to users for evaluation, feedback, advice and iteration. Rapid, low-fi prototyping will be used extensively to learn from users what works, where improvements might be needed and what does not work.

FIGURE 5.1   Chapter 5 outline

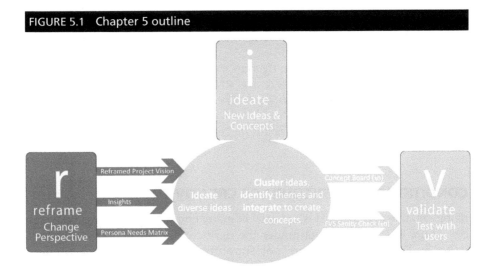

IN SEARCH OF HIGH-QUALITY IDEAS

Many innovation practitioners and commentators believe there is no short-age of innovation ideas in the world of business. Rita Gunther McGrath, Professor of Strategy and Innovation at Columbia Business School, has said: "Idea generation is seldom the problem! [...] to get breakthrough con-cepts, you really need some kind of framework" (Gunther McGrath, 2010). We agree.

Innovation ideas come in many shapes and sizes. Some are born out of frustration, necessity or insight in the course of everyday work. Alter-natively, other ideas are the outcome of reflective review and discussion with experts in the area, either informally over coffee or in the course of structured engagement for that purpose. Additionally, the role of seren-dipity in uncovering innovative ideas is well known, as in the discovery of penicillin by Alexander Fleming in 1928 through chance observation of an unintended effect in his laboratory. Fleming's predecessor biologist, Louis Pasteur, provides an incisive understanding of the roles of mindset and serendipity in new idea generation with his famous quote that "chance favours the prepared mind". Other innovation ideas arise from ongoing research and development programmes, some of which are clearly mission focused, while others are 'blue-skies' research.

We agree that ideas are plentiful in many businesses. However, we emphasise that what every business innovation project really needs are *high-quality ideas* that have a high probability of commercial success, in a timescale that suits the business. Happily, in the Ideate phase of the ARRIVE process, you are now in a good position to achieve this. You are working in a framework – as recommended by Gunther McGrath – that guarantees much greater probability of commercial success. The ARRIVE framework ensures that you are working on a solid platform for ideation with a clear vision of what you are aiming for, captured in the Reframed Project Vision.

By this stage, as in the Pasteur quote, your mind is prepared, you have a great understanding of your target users' problem space and you have clearly articulated insights relating to areas of discomfort, or dilemmas, and their root causes. You know their needs, the desirable outcomes, what they are trying to achieve, and you understand how they will evaluate success.

As you build your concept,
... the time you've spent on *rigorous research provides the bedrock*;
... the *Persona Needs form the central structure*; and
... your *insights ensure a rich, distinctive, compelling and complete offering*.
All the time,
... your *Reframed Project Vision is the compass* that guides you.

CONDITIONS THAT ENCOURAGE CREATIVITY

Being creative and focused is not easy in many circumstances. Everybody can contribute to idea generation under the right conditions but, mostly, people struggle. When conditions are not supportive, results can be disappoint-ing, and Teresa Amabile of the Harvard Business School has described the organisational conditions for creativity that managers can influence (Amabile, 1998). In summary, they are:

- **Challenge:** Match the job to the person so that the person is challenged but not overwhelmed.
- **Freedom:** Give people autonomy around process. Tell them which mountain to climb, but allow them freedom to decide how to climb it.
- **Resources:** Make sure the time, money and skills required are available.
- **Work-group:** Create groups with members who are mutually supportive, with diverse perspectives and backgrounds, and who share excitement about the project in hand.
- **Supervision:** Supervisors must encourage creative effort, not just outcome success.
- **Organisational support:** The organisation must reward creativity explicitly and mandate information sharing and collaboration.

GENERATING IDEAS

Teresa Amabile's conditions provide the necessary organisational backdrop for stimulating creativity in any project. Additionally, in this Ideate phase of the innovation process you should observe the following principles:

- **Focus on reframe/research outputs:** Do not allow the ideation exercises to stray too far or for too long from the *Reframed Project Vision*, *Insights* or core *Persona Needs*. Remember, these represent your evidence-based, hard-earned intellectual property designed to guide your process. The penalty for straying is usually wasted time and, maybe, outcome irrelevance. Occasionally, of course, some significant new insight on the problem might arise during solution ideation or prototype testing and this might warrant a revision. Certainly, capture this in a revised *Reframed Project Vision* or further detailed *insight* or *need*. However, be sure you understand how this relates to the original research data that created the *Reframed Project Vision*. Work hard to resolve apparent conflicts. While welcoming advances in your understanding and insights, be reluctant to ignore previous research data simply as a matter of convenience or to avoid solving the conundrum. If you thought it was valid and significant once, it probably remains so unless you can identify clearly why you were in error or why circumstances have changed.
- **Convert insights and needs into How Might We (HMW) statements:** Insights and need statements are, of course, centrally interesting. That is why you have isolated them for particular attention. When it comes to stimulating ideation, it is helpful to give them an imperative tone, or 'call to action', and you can achieve this by converting to the *How Might We* format. (We have done this already with insights recorded in our *Insight Notes* in the Reframe phase, chapter 4.) Of course, any language conversion implies a choice of emphasis, so there is likely to be more than one HMW question that may be derived from the need or insight.

Normally, however, a key, standout HMW statement is easily formulated. See Figure 5.2 for an example.

HMW

How Might We statements provide a powerful call to action for ideation sessions.

'How' implies there is a way, but we must seek it out.
'Might' implies multiple possibilities to be created.
'We' mandates collaborative working.

- **Assemble diverse groups:** Your project team has immersed itself in the subject area and has acquired expertise, but also perhaps some familiarity bias that warrants caution. Bring in outsiders from adjacent as well as distant domains to provide alternative perspectives and different thinking styles. Irrespective of what size or configuration the ideation group may have, ensure that everyone's voice is heard. In particular, you may need to guard against dominant senior managers having excessive influence.
- **Make the space right:** Successful ideation demands that team members immerse themselves in the topic for a sustained period of many hours per session, over multiple sessions. This cannot be achieved in an open office area, for example, with continuous interruption. Neither is it satisfactory for members to be popping in and out of involvement, responding to external queries, phone calls, etc. The ideation session ideally requires a dedicated room of comfortable size, plenty of writeable wall space, whiteboards and markers, Post-it notes, coffee

---

FIGURE 5.2 Formulating HMW statements from insights.

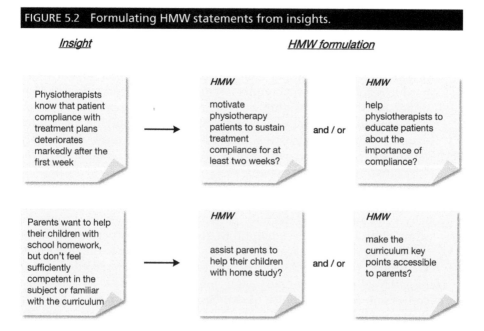

and water. It also needs a strong facilitator to keep the exercises and discussions moving along at a prompt and efficient pace, focused on the *Reframed Project Vision*, and being careful to capture all data that is created.

- **Diverge then converge:** Use multiple methods for generating a large quantity of ideas that address the research-derived *insights* and *needs*. Using a variety of methods will produce a rich diversity in resulting ideas. This is diverging, broadening the expanse of possible solutions. After these solution ideas have been generated they will be refined into themed clusters and moulded into draft concepts for preliminary evaluation. This is converging, i.e. reducing the array of options under consideration into a few, initially, and ultimately into just one integrated final concept.

Later in this chapter, we will show you our most used methods of idea generation. Remember, there is a great selection of methods available and new ones appear all the time. You should select the methods that suit you best and vary them regularly to keep freshness and novelty.

With the platform in place from your research it is now appropriate, for a while, to let go restraints and creatively explore a world of possibilities to solve your user's dilemmas in the most effective and compelling way. To start with, suspend judgement and switch to divergent mode, because a large quantity and diversity of ideas is the first goal. This is following the dictum of twice Nobel laureate Linus Pauling, who famously said: "If you want to have good ideas you must have many ideas". Remember, however, that you are now generating ideas with the aid of your guiding beacon, your new *Reframed Project Vision*. This is very different from wild, unfocused idea generation that characterises many ill-informed efforts!

Definitions

We use the term *idea* to refer to an undeveloped solution fragment, a thought or inspiration worth pursuing further to see if it has substance and potential as a full solution.

As ideas are clustered, extended and developed, we call the integrated value proposition a *concept* when it is phrased with emphasis towards the users' worldview (*desirability*), and communicates clearly the value proposition being offered.

When we add to the concept further information about market, technology, financial and strategic suitability, which are necessary for making a business case for internal-to-the-firm evaluation and decision, then we call this a *business proposal*.

When the concept and business model are fleshed out and capable of delivering value to the customer as intended, then we call this a *product* (or *offering*). In its first usefully deployable configuration, it is an *MVP* (minimum viable product).

We emphasise these distinctions now to avoid confusion as we progress in particular through the Validate and Exploit chapters later.

BUILDING CONCEPTS

When you have (divergently) accumulated perhaps hundreds of 'idea-frag-ments', change tack to converge, i.e. start to cluster and put shape on the ideas. While some ideas may adopt a bird's-eye or strategic view, many will be detail-oriented, focused on a specific insight or need. To create a concept, you must zoom out and take a holistic, integrative perspective. Team discussions, dialectical criticism, critical thinking and reflection all play a role in filtering the mass of ideas to a subset with greater relevance and potential. While it is great, at first, to have many ideas, the user and other stakeholders only can respond to and get value from a coherent, cohesive, integrated value proposi-tion in the form of a concept.

A concept is a high-level, clearly themed description of your user value proposition, designed to communicate the value proposition's essential ele-ments clearly and without clutter of detail. The early concept's main purpose is to aid communication with the users and other stakeholders in order to elicit feedback about the value offering being proposed and advice on how to improve it. It also assists in establishing a common vision with innovation team colleagues, as well as diverse business functions within the firm and partners, and forms a platform for integrating good ideas and developing them further. Above all, a concept helps answer questions about the *desirability* of a proposed innovation. Later, if it survives evaluating scrutiny, the concept evolves into a *business proposal* that includes relevant aspects of the viability, feasibility and suitability dimensions.

Concept building is one of the most exciting and satisfying stages of an innovation project. It brings together all the fruits of research, reframing and ideation into integrated themes to propose new complete solutions. Later in this chapter, we will show you how to represent a concept clearly in the form of a concept board. The concept board is a visual and accessible overview of key features and parameters, with a zoomed-out perspective that removes dis-traction by less important details. The concept board is the user-facing expres-sion of the ideation and concept building outcome. For the innovation team itself, of course, all outcomes are captured in the finer detail of a design file of research notes, sketches, calculations, test results, etc.

Early concepts focus on the core value proposition, which aims to sat-isfy core needs and responds to the most impactful insights. Later concepts evolve progressively to incorporate richer functionality and more complete value offering. The evolution continues in response to new learning and the team continues to cater for additional secondary needs and insights, as many as it may sensibly accommodate. At early stages of concept development, you might expect to have tens of skeletal concepts, which are realised as sketches and simple cardboard models or equivalent. Often the innovation team co-creates these in collaboration with prospective users.

Eventually, through filtering, iterating and refining the earlier models, two or three shortlist concepts emerge, which address the core need of the *Reframed Project Vision* from different perspectives and respond to partially different subsets of secondary needs and insights. The innovation team will subject these shortlisted concepts to more detailed and rigorous assumption testing and validation, as in chapter 6, Validate. The ultimate goal, after vali-dation, is to converge on a final integrated 'candidate' concept that you will develop and execute to deployment.

FIGURE 5.3 Concept board layouts.

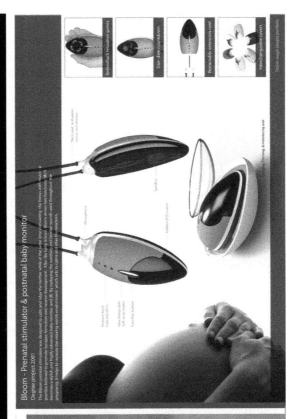

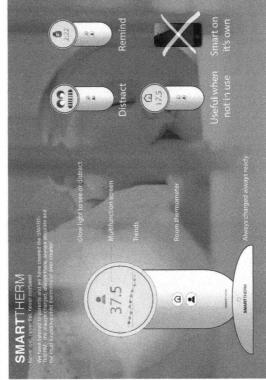

Converging from hundreds of ideas to tens of early concepts and finally to one integrated concept is like a co-creative improvised jazz session. It eventually achieves a climax of distinctive fluency, or flow, which transcends individual contributions. Here, multiple users, stakeholders and other co-creators engage in rapid-cycle demonstration, testing, feedback and further ideation to converge on the 'ideal' solution.

A NOTE ABOUT ENGAGING THE USER IN CONCEPT DEVELOPMENT

Some people describe an innovation team's involvement with the user as being like an accordion – it entails repeated in-out-in-out motions to produce the music. The Audit phase of the ARRIVE process begins with desk research before the Research phase requires intensive engagement with the user, gathering information and building understanding. After this, the Reframe is more introverting and internally creative, when the team intensively ruminates over the information gathered earlier, absorbs its meaning and unravels the hidden insights.

The next activity, idea generation, which we have just described and forms the first half of our Ideate phase, is done by the team alone or with selected guests. Divergent idea generation, hand in hand with critical reflection and dialectic discussion by the team, produces a filtered set of ideas with potential to build upon.

Now, the upcoming concept-building phase requires another re-engagement with the users. Their inputs and feedback help the team make the right choices regarding which ideas to drop, which ones to develop, in which direction to develop them, and the ecosystem that must support them. The quicker, sooner and more frequently you do that, the better! As a rule, it is always better to engage with the users as far as possible within the environment where they will use the proposed solution, so that the responses will be most genuine and reliable. There, drawing on the environmental inspiration, you can more fruitfully interact with users to discuss sketches, models and enactments, or tell narratives that illustrate and elaborate on the way your concept is developing. The feedback that comes in the forms of verbal or tacit expressions and subtler, even unconscious, behaviours greatly enriches the concept development process.

FIGURE 5.4 Interaction between innovation team and user is like an accordion.

FOCUSING ON FEW, DIVERSE CONCEPTS

Throughout the concept development phase in most commercial organisations, you will not have the luxury of time and money to carry forward multiple options for too long. As concepts develop, assumption testing and prototyping become quickly more complex, time-consuming and expensive, and bosses become impatient. Conversely, you should not converge too soon on a single concept, and thereby rule out other ideas before the team has properly explored them.

What should you do? Sometimes the organisation's management may drive this issue beyond the control of an innovation team leader and 'force' a decision. Alternatively, some exercises such as Google Venture's Design Sprint explicitly prescribe appointment of a 'Decider' who can make a binding decision in case of prolonged team uncertainty over choosing alternative options (Knapp, 2015). At the end of the day, it is a judgement call the innovation team leader must make based on the team's and the organisation's circumstances. Our advice is simple: try not to give in too easily to the inevitable pressures that seek to reduce your options too early to a lone concept. Keep multiple options and hence learning opportunities open for as long as your circumstances reasonably allow.

There is a big difference between being compelled by circumstances to cull your options, as described above, on the one hand and, on the other hand, being simply unable to come up with alternatives. Sometimes it is difficult to get your attention away from an anchor idea once it has established itself in the team's collective psyche. We use a method that we call M3 to help avoid this and to work towards producing intermediate stage concepts for three scenarios. The scenarios are categorised loosely in relation to the parameter 'effort to deployment':

Mild concept	Provides incremental improvement to the users' dilemma and is relatively easily implemented, equivalent to 3–9 months' effort to deployment with in-house or readily available resources.	
Moderate concept	Provides worthy and substantial improvement to the customer and significant differentiation to you as supplier. Entails a substantial project, *moderately* stretch goals but with no foreseeably insurmountable difficulties, modest risk, equivalent to 12–24 months' effort to full deployment.	
Mad concept	(Of course, we use 'mad' in the young person's positive sense of 'wild' or 'out there'!) This provides the almost perfect satisfaction of the *Reframed Project Vision*, without compromise, addresses all insight dilemmas and identified needs, and is game changing for the users and the broader market. Compelling *desirability* is its clear focus. Its feasibility is constrained only by what would be 5–10 years' development effort-to-deployment in a large, well-resourced multinational or priority, government research project. Viability and strategic suitability matches are still uncertain or unknown because of the scale of the ambition.	

The three scenarios above are analogous to the *three horizons of growth* beloved of strategic planners.

We recommend that the three scenario prototypes are developed in the sequence shown, starting with 'mild'. It seems to work better for most people when they get the easier mild material out of their heads first and they can then apply themselves with better concentration to the more adventurous concepts.

Of course, be sure continuously to review how your *Reframed Project Vision* is satisfied by the concepts you create, and refine the concepts (or the *Vision*, cautiously) accordingly.

FROM CONCEPTS TO BUSINESS PROPOSAL

When you iterate through multiple evolving concepts, all the time you are making intuitive evaluations as to which you think may be 'better', i.e. more impactful for the user as well as more viable, feasible and strategically suitable for your firm to deploy.

At the earlier stages, as we have said before, we encourage you not to let concerns of *viability, feasibility* and *suitability* dominate your thinking too much. It is easy to let these more practical issues drag you down into bland conservatism and inhibit your bold moves. Instead, give priority to the imperative of desirability and let your ambitions soar so that you may uncover something really compelling for the user before you tether yourself back to practicality with the other three factors of innovation.

Nevertheless, while desirability and user validation are the central focus of early stage concept development, clearly it is prudent to keep a strategic eye on the other factors of innovation to ensure the concepts stay within reasonable bounds. As concepts progressively gain more definition, regularly track reasonableness or *sanity* of a concept by using the *FVS Sanity Index*, as shown in Figure 5.5. This provides a coarse, but useful, go/no-go evaluation or sanity check to give greater confidence in a concept's realisation potential or time-to-deployment, or to highlight other areas that may need to be investigated before proceeding to further development.

THE BUSINESS PROPOSAL – GETTING READY FOR BROADER VALIDATION AND DEPLOYMENT

All the while with concept development you are juggling between concepts to kill and concepts to bring forward. You cannot evaluate and filter concepts with a simple algorithm. It depends on the type of project, the scope of project and the stage of its development.

Eventually, you will proceed to the next phase of the innovation process, Validate (chapter 6), with a shortlist of two or three *desirability*-prioritised concepts. You must extend each of these to embrace a wider range of business-relevant considerations such as business model, market size, competitive positioning, expected margin, capital investment requirement, technology platform, strategic alignment and many other *feasibility–viability–suitability* factors.

As the *concepts* evolve into *business proposals*, they must be amenable for senior management to evaluate and compare them, from a business strategic perspective. The key components of a business proposal summary that we use to accompany the *Product Concept* Board (B) are the *Lean Canvas (Business Model)*

FIGURE 5.5 FVS Sanity Index template – checking concepts against feasibility, viability and suitability.

FVS Sanity Index
(sanity checking Feasibility-Viability-Suitability factors of early stage concepts)
Score < 15 in any category = serious problem!
Score >20 in all categories = looking good ...... if the user wants it!

Concept name:

Feasibility

Are the likely implementation technologies ...		What is the level of execution risk?		What is the time from development start to deployment readiness?	
New to the world?	1	Very high	1	> 24 months	1
Stretch for the firm?	3	High	3	<18 months	3
New to the firm?	6	Moderate	6	< 12 months	6
Routine for the firm?	10	Low	10	weeks	10

Comments

Feasibility score (/30)

Viability

How scalable is the concept?		What is the margin possibility?		Is the required investment ...	
Not much – very specific	1	Tight	1	Special?	1
Moderate	3	Moderate	3	Large?	3
Good potential	6	High	6	Routine?	6
Exceptional	10	Exceptional	10	Minimal?	10

Comments

Viability score (/30)

Strategic Suitability

How well is this concept aligned with our current mission and strategy?		How great are the organisational changes necessary?		Do we have/will we get enough high-level support in the firm?	
Not clearly aligned	1	New division/company	1	Not clear, not explored	1
Not clear	3	New team	3	Broad team support	3
Aligned, but new activtity	6	Moderate	6	High-level champion	6
Fully, clearly	10	None	10	Clear, strategic, CEO level	10

Comments

Suitability score (/30)

FIGURE 5.6 Emphasis of early stage concepts and later stage business proposals.

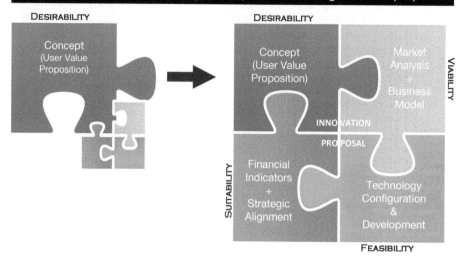

and *Business Case Dashboard*, which we introduce in the next chapters 6 and 7. Additionally, and integrally, a *business proposal* must highlight key assumptions that still await validation, and questions that are so far unresolved. The role of Validate (chapter 6) is to de-risk these assumptions as much as possible, through testing before launch. The Execute phase described in chapter 7 carries on with this testing and iteration, but under the even more intense pressure of requiring real performance with real, paying customers, in the marketplace.

In the next chapter, Validate, we will discuss in more detail assumptions mapping and testing.

The role of assumptions

A concept or business proposal's prospect for market success is evaluated by considering all elements as an integrated interacting system. However, all through the development up to the final launch, the proposal's key premises are a mixture of verified facts, uncertainties, creative leaps and unknowns. Rigorous certainty in all parameters is impossible until after deployment. Even then, it is often only in hindsight that proper evaluation of a business proposal's success can be established.

Design thinking's approach to this dilemma is to make explicit the filling-in of unknown spaces with educated estimates or *assumptions*, and then to proceed with selected concepts or business proposals by verifying (validating) or amending the assumptions through repetitive prototyping and testing. Initial assumptions and estimates are refined through testing and analysis to become evidenced facts, as closely as possible.

In this manner, assumptions are like oil that allows the innovation engine to move forward. Perhaps a better analogy is that *assumptions in an innovation system are like credit in a financial system*. Both are educated guesses about the future that allow action to proceed, albeit always with some risk.

A NOTE ABOUT CONCEPT RADICALNESS, UNCERTAINTY AND TECHNOLOGY

Sometimes, project briefs are clearly directive towards incremental change or radical change or some specific level in between. At other times, the degree of innovation emerges in response to the data uncovered in Audit or Research phases, taking into account a negotiation with the firm around time-to-completion and other considerations. Often, there is a presumed exclusive correlation between radicalness or risk of an innovation solution on the one hand and technological uncertainty or time to deployment, alone, on the other hand. Such a correlation does not always apply. It is not only technology that is risky.

Many radical, paradigm-changing innovations are highly risky and yet technologically straightforward. For these, the risk may be concerned with correctly anticipating how users' behaviours will change or adapt to the new business model or adjusted ecosystem proposed by the innovation, as Adner and Kapoor (2016) have written about recently.

The emergence of low-cost airlines such as Ryanair and Southwest Airlines more than a decade ago provides an example where uncertainty, risk and potential reward were associated with radically new business models rather than technology. Prior to this, operating norms of the industry included multi-stage flight booking, complimentary in-flight refreshments and travel agents as booking intermediaries, for example. The new business models eschewed these and other 'normal' components of business in the expectation (hope) that resulting lower cost fares would cause travellers to change their perception of air travel from an occasional luxury to a commonplace activity and so change travel behaviour.

Happily, so it transpired, and these airlines have built over a decade of success from it. The technological uncertainty was low, the organisational and business model set-up required skill and time, but the planned outcomes were clear and reliably expected. However, the uncertainty regarding users' behavioural response was high.

On the other hand, new technology development is always fraught with uncertainty of performance and time-to-deployment, and a new technology must go through many stages of development from highly uncertain discovery to fairly certain (re-)deployment. The US National Aeronautical and Space Administration (NASA) was the first to codify the difficulties associated with technology developments in the early 1970s when it developed the concept of the Technology Readiness Level (TRL). Later, in the 1980s, NASA defined the TRL scale formally for vendor and other contracting purposes. Since then, many other organisations and industry sectors have adapted the scale to suit their own contexts. See table 5.1 for the European Commission TRLs, which it defined for use in its Horizon 2020 research programmes within the last decade (European Commission, 2014).

For most organisations, TRL 6 or higher is the stage when a technology may be considered for an imminently commercial innovation project and the risk due to uncertainty is acceptable. Below TRL 6, the technology is still in the form of an engineering or scientific research project, where risk is higher and it is not suitable for near-term implementation. At TRL 3 and above, it may be considered for longer-term (technologically radical) concept possibilities. From TRL 1, it is useful for stimulating 'moonshot' possibilities or dreams, as in the 'mad' concepts that we described in the previous section.

Table 5.1 Technology readiness levels.

Technology Readiness Level	Description
TRL 9	actual system proven in operational environment
TRL 8	system complete and qualified
TRL 7	system prototype demonstrated in operational environment
TRL 6	technology demonstrated in relevant environment
TRL 5	technology validated in relevant environment
TRL 4	technology validated in lab
TRL 3	experimental proof of concept
TRL 2	technology concept formulated
TRL 1	basic principles observed

Source: European Commission (G), 2014.

METHODS IN THIS CHAPTER

As we have said above, the first objective in idea generation is to maximise, with clear focus on the problem being addressed, the quantity and diversity of ideas that are available for consideration. There is no shortage of techniques, and new variations appear all the time. We have chosen a sample for presentation in this chapter, which we have found give energy and inspiration to our ideation and conceptualisation sessions, for both executives and university classes. You may have your own favourites and, of course, you should use them with or instead of those given here.

For generating multiple solution ideas, we introduce the following methods:

- **Warm-up exercise:** The creative 'right-brain' is like a muscle that does not get as much use as it should. It needs practice. It also needs to be warmed up before put to most productive use.
- **Free-flow brainstorming:** At the start of every idea generation session, participants usually have a store of ideas already in their minds that have developed throughout the prior period of research and experience. While valuable, they may also act as blockages to more expansive thoughts. Hence, the first activity should always be a free-flow brainstorm, which captures stored ideas and frees the mind for new thoughts.
- **Through-the-Eyes:** Seeing the situation through the eyes of an archetypical character or brand helps to enable a free-flowing narrative of how the archetype would address the predicament in a particular way.
- **Inversion (or Bad Idea/Good Idea):** Perversely, it is often easier to think of an obviously bad idea than a good one. At least, it usually seems they are more plentiful and emerge more quickly. Once listed, the bad idea may be inverted to see if something valuable may be found from its opposite.
- **Metaphor:** Metaphors are powerful means to stimulate new perspectives and narratives for any situation, to extend the range of thought and to suggest possible consequences.

For developing concepts, we introduce the following methods and templates:

- **Concept synthesis:** The many ideas that you have derived from a process of stimulated divergent thinking will be varied and maybe overwhelmingly so. We show you how to make sense of all the ideas and to converge on two or three coherent concepts that you and the team will further test and develop.
- **Concept boards:** A clear presentation of the most important elements of a concept. It shows how they integrate to provide a compelling value-enhancing proposal for the identified user and is invaluable as an aid to thought, communication and gathering feedback.
- **FVS Sanity Index:** This template is a tabulation of common elements of the feasibility, viability and suitability factors, which allows quick evaluation and go/no-go filtering of early stage concepts.

OUTPUTS AND TEMPLATES

The outputs that you take from the Ideation phase are clearly articulated concepts – most likely two or three, maybe more – which now require detailed validation. In addition to the broad learning and the creative output assets that your team has built up, the specific outputs are concept definitions in the form of:

- **Concept Board (A)** (multiple)
- **FVS Sanity Index** completed template (multiple)

Ideate | methods

a audit
What is known?

r research
Find out more

r reframe
Change Perspective

i ideate
New Ideas & Concepts

v validate
Test with users

e execute
Develop, deploy & scale

 1

 2

 3

WARM-UP EXERCISE
The creative 'right-brain' is like a muscle that does not get much use. It needs practice and warm-up before working effectively.

FREE FLOW BRAINSTORMING
Capture stored ideas and free the mind for new thoughts.

THROUGH THE EYES
See the situation through the eyes of an archetypal character or brand.

 4

 5

BAD IDEA / GOOD IDEA
Invert bad ideas to find something valuable.

METAPHOR
Metaphors stimulate new perspectives and narratives, to extend the range of thought and to suggest possible consequences.

 6

 7

 8

CONCEPT SYNTHESIS
Converge on two or three coherent concepts that you and the team will further test and develop.

CONCEPT BOARD (A)
A clear presentation of the most important elements of a concept.

FVS SANITY INDEX
A checklist for regular evaluation of common elements of Feasibility, Viability, Suitability dimensions.

Inspired by the insights from Research and armed with a Reframed Project Vision and Persona Needs Matrix, generate multiple solution ideas and weave them into rich concepts that will delight the user.

WARM-UP EXERCISE method   1/4

Prepare the team for high-performance creativity.

"Let's do a warm-up exercise" – six words that can bring a collective moan from any group. The warm-up at the beginning of a meeting is often seen as a silly waste of time and, sometimes, dignity. Many people do not see a purpose. However, warm-up exercises in innovation sessions serve a similar role to warming up for physical exercise – they prepare the mind muscles for what is coming. It is well proven in many studies that a good warm-up prepares participants and team for the type of thinking and mindset that is necessary for innovation to work – curiosity, open-mindedness, creativity and even a little vulnerability.

The second important function of a good warm-up is to prepare the group for something different outside of day-to-day operations, to get comfortable thinking a little differently and opening up, sharing and collaborating.

There are hundreds of different exercises to choose from that will warm-up a group and we encourage you to look around and find approaches that fit what you are trying to achieve in each session. Here, we have selected one exercise that we find ticks a number of important attributes for a successful session. We call it 'forced connections'.

In the forced connections method, participants are presented with random objects, places, things, etc. and they are asked to combine them into new value propositions – the more interesting and creative the better! This method activates an essential component of creativity and innovation – combining seemingly disparate things to create something new.

As an example, if you were given the cards 'cat' and 'passport', you might come up with something simple like a new travel passport for animals, or, more interestingly, perhaps there could be a holiday hotel for animals that spoils and relaxes the animal when you are on holidays.

Step by step

1. **Prepare the exercise**
   First, collect a number of images of random things – the more interesting and diverse the better! They could be basic and obvious things like a kettle, a car, birthday cards, bread, etc., but you can make it more interesting by adding images of things like a passport, tinder, a pile of cash or even a government. It really does not matter what you choose; the point of the exercise is to stretch the group's thinking.

2. **Let the participants know why**
   After you have welcomed the group to the meeting, let them know that before you begin you are going to do a warm-up exercise. To reassure them, explain the importance of a warm-up and how it compares to warming up for physical exercise. If you have not done so already, it is a good idea to mix the participants into groups.

3. **The format**

   Establish groups of three or four people. Print the images on A4 paper and distribute to all teams. To make it more fun, have them pick pairs of images out of a hat. This always gets a chuckle, especially when someone picks out a particularly difficult or funny pairing! To get the whole group involved together and to make it somewhat competitive, you can present the random images side by side on a projected slide.

4. **Let imaginations fly for 15 minutes**

   Instruct the group to spend five minutes per combination trying to come up with a unique value proposition in the form of a new product, service, experience, business model, etc. This exercise is more concerned with stretching imagination and credulity than creating realistic and sober ideas. Let the teams share the results with the whole group. Give kudos for humour, creativity and innovation.

WARM-UP EXERCISE   template   2/4

FIGURE 5.7   Forced connections template.

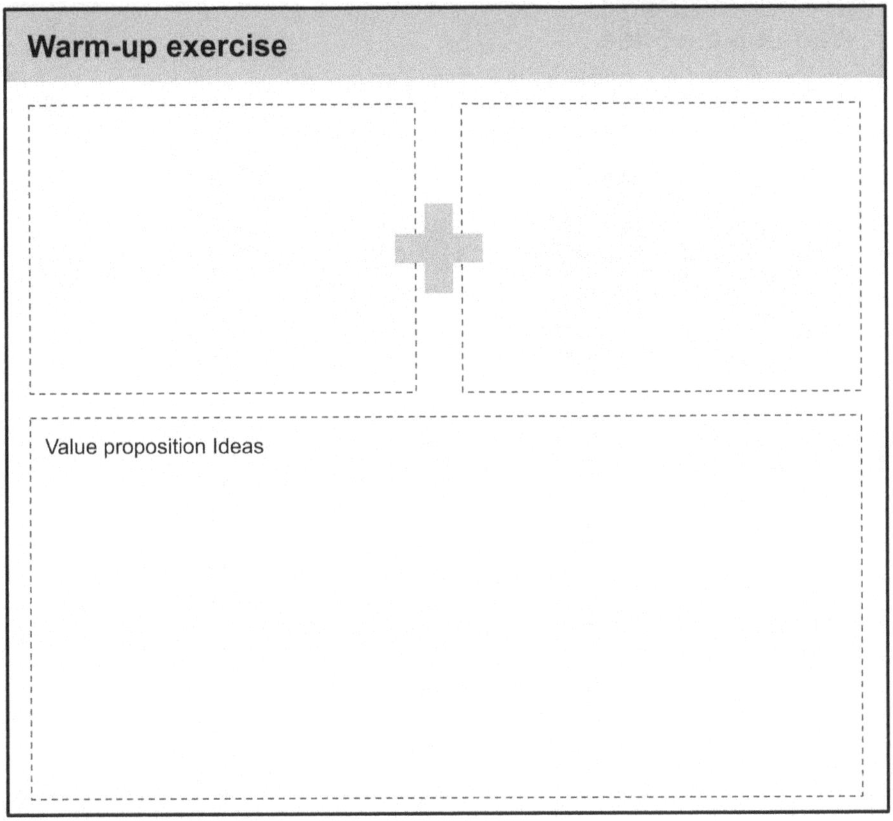

Warm-up exercise

Value proposition Ideas

FIGURE 5.8 Forced connections example.

Warm-up exercise

Value proposition Ideas

A special hotel and spa for your pets

A service where you can officially adopt a pet into your family.

Cat minding service for international visitors to a country

A 'nightclub' for well-bred animals. Proven by a pet passport.

A service where you can get your pets photographed, chipped and certified

WARM-UP EXERCISE faqs 4/4

Q It feels like a waste of time coming up with silly and infeasible ideas. How can I justify it to sceptical team members, for example engineers and accountants?

A Innovation is a serious pursuit but the route to excellence cannot be paved only with in-depth, serious consideration of everything. To find the optimum path requires open-minded exploration in many directions. We think of it more as serious play, where fun and a little silliness loosen our thinking to allow for more creative, yet serious, ideas.

Q If someone cannot come up with many or any ideas, does that mean she is not creative?

A That would be a very harsh judgement! Most likely, it means the person is not used to thinking abductively, making leaps of imagination that cannot, a priori, be justified. Alternatively, it may mean the person is shy and afraid of what might be thought about her for saying something silly. But, this is the point of the warm-up, to start to overcome these barriers. If the lack of confidence is deep or the experience with abductive thinking is shallow, then it may require more than a warm-up. Remind everybody that practice always helps.

Q Can we use this technique in brainstorming or is it just a warm-up tool?

A Absolutely, you can! Combining things is the essence of creativity and, at its simplest, innovation is just bringing disparate things together …. think wheels and suitcases, air and trainers, hotels and ordinary people with spare rooms, etc. We think this is a great warm-up exercise, because it can be done without any preparation and is immediately accessible to all. When you start applying it to your own challenges it can unlock some great ideas, too.

When you seek solutions to a problem, group brainstorming focuses attention and encourages quick, spontaneous ideas.

Brainstorming is a creative skill that you use regularly throughout design, innovation and problem-solving projects. It provides a safe place for a group of people to share good and bad, simple and wild ideas, and to be heard without judgement. It also serves to help people "get out of their head" ideas that are already there and thus be ready for expanding the pile through more structured approaches.

The most important thing about brainstorming is that there is little point doing it unless you have a clear goal, challenge or problem you want to tackle. Before you start, be sure to put effort into defining challenges to be tackled and clearly articulate these at the beginning and throughout the session.

A further important point is to note that without rules and structure the session will not deliver value. Instead, it becomes a disorganised free-for-all with the loudest in the room getting their voices heard and good ideas getting lost. The innovation consultancy IDEO has done great work in this area and has identified seven 'ground rules' for a successful brainstorming session. These rules can be found on the following page.

While there are ground rules, their sole purpose is to ensure that the session leads to productive ideas. A brainstorming session must be fun, high energy and lead to in-depth, exploratory conversations. To make this happen, it is almost mandatory to have a nice airy room, lots of space, coffee, water, sugar and energy-giving snacks.

Finally, it is normal that ideas tend to slow down after a while and there are many techniques that extend the flow of ideas and change how you might think about the challenge. We cover some of our favourites in the following sections.

Step by step

1. **Choose the goal of your brainstorm session**
   What are you trying to accomplish with your brainstorm session? It is important to represent this with well-defined question/s.
2. **Allow imagination and big thinking**
   Encourage the group to go wild and think outside of the box. Good brainstorming sessions encourage radical ideas. It is easier to scale back extravagant ideas than make small ideas into something bigger.
3. **Build on the ideas of others**
   The power of brainstorming lies in the group's ability to build on and extend others' ideas, like the flow of a jazz band. The best sessions will have members who are comfortable enough to put silly or far-fetched ideas out, as these often lead to powerful suggestions. British actor John Cleese once called these the "intermediate impossibles".

4. **Do not evaluate**

A common barrier to a successful brainstorming session is the normal human tendency to evaluate and critique ideas too early. The best ideas from a brainstorming session, or any idea generation technique, come when the group builds up rather than tears down. The time will come later to critique, filter and iterate 'crazy' ideas into something viable.

5. **Go for quantity**

Linus Pauling said, "the best way to have a good idea is to have lots of ideas". Push the group hard by setting a stretch goal, e.g. each team should try to come up with 50 ideas in 20 minutes, or 30 in 10! The time pressure discourages premature evaluation.

6. **Be visual**

Communicating and remembering are equally as important as creating the ideas. Use visuals as much as words. Where possible, give the idea a name or title so that it conveys the essence and is more easily remembered and referenced.

BRAINSTORMING template   2/4

Table 5.2 IDEO's Brainstorming rules (ideou.com/pages/brainstorming).

1.	Defer judgement
2.	Encourage wild ideas
3.	Build on the ideas of others
4.	Stay focused on the topic
5.	One conversation at a time
6.	Be visual
7.	Go for quantity

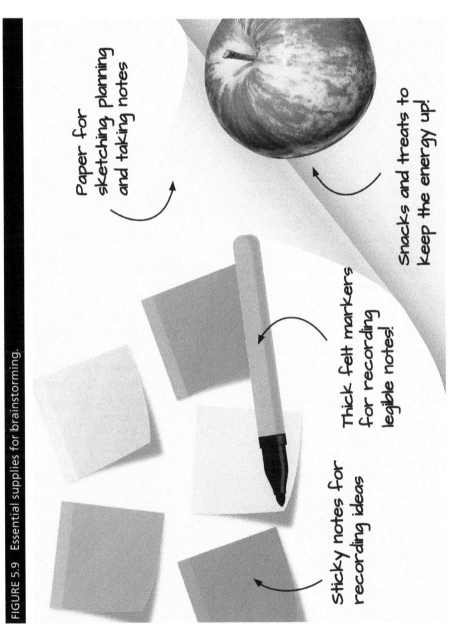

FIGURE 5.9 Essential supplies for brainstorming.

Paper for sketching planning and taking notes

Snacks and treats to keep the energy up!

Thick felt markers for recording legible notes!

Sticky notes for recording ideas

Q We have tried brainstorming before. How can we stop the group constantly going off on tangents?

A We cannot overemphasise how important it is to have a well-defined challenge from the off. With a well-defined challenge you have something to keep the group on track, to pull them back if they stray.

It is also important to have a good facilitator to keep the group on target. The facilitator must be fluently familiar with the various techniques used. In addition, an experienced facilitator will understand intuitively when to rein in a conversation that is going astray and when to allow a conversation to continue if it appears to be heading in a fruitful direction.

Q How many people should be present for a good brainstorming session?

A There is no magic number. More than three people makes for more interesting and diverse conversations. More is better and will produce more and more-diverse ideas, but to keep the session manageable and on-track, you must break up larger groups into smaller operating sub-groups or teams of, say, three to six members.

Q After an initial burst of ideas, we often seem to run out of steam. Apart from trying new brainstorming techniques, are there other ways of getting more out of a group?

A Coming up with new ideas is very energy consuming, so it's not surprising if a group slows down after an initial blast. Apart from the various techniques that we discuss in this chapter, there are other tricks you may use.

Try the following. Ask the group to number each idea and try to come up with 50 ideas in 15 minutes. Or, you could have a 'round robin', so that when one person comes up with an idea the next person must build upon it.

Whatever method you use, be sure to persevere past the initial valley. It always happens that a group gets a second wind that builds on the first.

THROUGH-THE-EYES method   1/4

This is a creative technique that challenges a group to take a new perspective on a problem by adopting another's problem-solving personality.

It is an excellent tool for extending the flow of ideas and unlocking creative solutions to an identified challenge. In traditional brainstorming, an individual is often bound to a paradigm and logic that is full of assumptions about how the world works, how to go about solving the problem and even the nature of the problem.

In order to extend the range of emerging ideas, a good device is to consider it with a fresh worldview paradigm, understanding and experience. To support this, the Through-the-Eyes technique asks participants to temporarily adopt the personality and worldview of a well-known character, and to look at the challenge through the mind's eyes of that person.

It can be surprising how role playing different characters unlocks very different ideas for the same challenge. Whether your objective is to create an outcome with a great user experience, something that will reignite a market, be cool and sophisticated or be completely new to the world, there is a character in the world that can help.

For example, if you are trying to get the participants to create a magical experience in a complicated system, we might ask them to view the challenge through the eyes of Walt Disney. If you are looking for a common-sense approach to a social challenge, you might ask participants to see the challenge through the eyes of a mother. Some of the characters we find particularly powerful are included on the next page. Of course, we encourage you to add your own.

Step by step

1. **Identify your characters**
   In advance of the workshop we like to spend time identifying four or five characters with personality attributes that align with the outcome we are trying to achieve. Very recognisable 'celebrity' figures work well, but you can also use brands or companies that have matching qualities.

2. **Identify the challenge**
   As always in ideation, begin the session with a clear challenge in mind and remind the team of it regularly.

3. **Explain the exercise**
   Let participants know that in this exercise they will adopt a particular role, perhaps multiple roles in succession. Each time, they will come up with solution ideas for the challenge from the given character's perspective. Note well: it is the 'character' who is solving the challenge for you! It is not that you are solving a problem for the character – sometimes a mistaken interpretation.

   Remind participants of the principal attributes of each character being adopted, in order to help them get into and stay in the role.

4. **Begin multiple ten-minute sessions**

Project your characters on screen one at a time, for about ten minutes each, allowing the team to give their complete focus to that character alone. You can simply project an image of the person with attribute key words. Or, a short video clip can also work well. Throughout the ten minutes, you may choose to mention some examples of how that person has taken on a challenge, something they have done, or perhaps a relevant quote.

5. **Share**

After each character, ask the team to share interesting ideas. This helps stimulate new thoughts immediately and for following characters.

THROUGH-THE-EYES template

2/4

FIGURE 5.10 Through-the-Eyes candidate characters.

FIGURE 5.11 Through-the-Eyes example.

DAVID BOWIE

If **David Bowie** was in charge of your project, how would he solve this challenge?

Attributes

enigmatic   self-assured
showman   energetic
artistic

IDEAS

Q We notice that some people find it difficult to adopt the character's persona. Is there a way to make it easier?

A Perhaps they over-think it. Emphasise that this exercise is not meant to be a 'unique solution-finding' one. It is an opening-up session, where everyone suspends reality for *ten minutes* for each character. The immediate objective is simply to fill the table/board/wall with ideas. Evaluation and rationalisation of ideas come later and, without doubt, the final solution will be better when there is a wide variety of options to start with.

This point reinforces, also, the importance of warm-up. Some need it more than others, in order to loosen the imagination.

Ask participants to think about things the character has created and the personality attributes that helped them do it. For example, Walt Disney created a magical world with a reputation that travels worldwide. He created Disneyland as a destination. As a person, Walt was all about magic, make-believe and storytelling. In your project, could you turn your headquarters, store or factory into a destination? Can you add a little magic to what you do or how you communicate? It's worth ten minutes to think about it!

Q Can I do this exercise 'through the eyes' of 'normal' people like our CEO, a customer or a competitor?

A You can try! This technique is designed to channel and emulate the approaches of individuals or entities with strong personalities, viewpoints or ways of working, thinking and doing. When such people have done something big, impactful, brave or exciting, you may be able to take inspiration from this and apply some of it to your own challenge.

Such people are found in the public domain among inventors, actors, sportspeople, business or political leaders. They are in the public domain, often, because of their eccentric characteristics and exceptional performances and thus they are easy to identify and emulate.

If you – and your team – know such a person, then that could work. However, in our experience it is unusual for the whole team to identify with such a person, especially if that person is not already involved in the firm or project. Remember you are looking for 'non-standard' inspiration.

Deliberately look for the worst ideas in order to transform them into something good.

When we think of brainstorming we most likely think of a group of people trying to come up with the best ideas as solutions to challenges. Done right this can lead to clever, sometimes breakthrough ideas. However, when ideas dry up and our natural bias or assumptions take over, idea flow can become constrained. Additionally, some people have a natural reluctance to let loose in public and are afraid to suggest something that may make them appear stupid. It is probable in these circumstances that there is a lot more 'fuel in the tank' that might be left unused.

The Bad Idea/Good Idea technique is an approach to get to those hard-to-access ideas and to bring further new perspectives to the challenge.

While it may sound counterintuitive at first to encourage the group to look for bad ideas, the approach helps to relieve anxieties of some participants and gives everyone permission to put out crazy ideas and think about the challenge in a different way. This is a major benefit of the technique.

The increased psychological freedom brings a new energy to the session.

In addition, looking for bad ideas encourages participants to navigate around firmly held assumptions and to reveal, perhaps unintentionally, where these assumptions are not soundly based. In our experience, it is not uncommon for at least one-third of 'bad' ideas to hold inspiration for potentially good ideas worthy of further consideration.

Step by step

1. **Brief the group**
   Remind the group of the specific challenge before them. They should think about it in a way that is polar opposite to normal. Rather than good ideas, look for bad, terrible, illegal, unethical, even disgusting ones. A suitable bad idea is one that (from present perspectives) would definitely hinder prospects of a successful outcome. Be extreme!
2. **List ten bad ideas**
   Working in teams of 4 or 5, the teams come up with ten really bad ideas and list these on the left-hand column of the template provided. This may be on a flip chart or a whiteboard. We suggest allocating ten minutes for this part of the exercise.
3. **Show them the way**
   This exercise may feel unnatural for some participants and some simply may not get it! At an early point, you might highlight an example from one team, or throw in one or two of your own bad ideas.
4. **Identify attributes**
   When the teams have identified ten bad ideas, ask them to go through each and, on the same template, identify two or three key attributes that make it bad. For example, a terrible idea for our challenge 'Reducing speeding' might be to 'incentivise high speeds'. Two attributes of this

might be 1) rewarding bad behaviour, and 2) understanding the drivers' desire to speed. Allow ten minutes.

5. **Invert to good ideas**

Finally, ask the participants to generate possibly good ideas by inverting the sense of the bad ones or being inspired by them. It does not have to be an exact inversion, just anything that is brought to mind that might be at least somewhat feasible. For instance, 'reward bad behaviour' could become 'reward good behaviour', where people who obey speed limits get put into a lottery. Allow 15 minutes.

BAD IDEA/GOOD IDEA   template   2/4

FIGURE 5.12   Bad Idea/Good Idea template.

Challenge:	
BAD IDEAS	**GOOD IDEAS**
1 Attributes:	1
2 Attributes:	2
3 Attributes:	3
4 Attributes:	4
5 Attributes:	5

BAD IDEA/GOOD IDEA   example   3/4

FIGURE 5.13   Bad Idea/Good Idea example.

Challenge: How might we reduce speeding amongst young men, particularly in cities.

BAD IDEAS	GOOD IDEAS
1 Give rewards to drivers for going fast Attributes: Rewards. Understand what people want.	1 Reward drivers for obeying the speed limit. Perhaps a lotto system based on licence plates,
2 Clear the streets of pedestrians so cars can go fast. Attributes: Make streets safer. Remove all obstacles to cars.	2 Identify areas where cars usually speed and make them safe for faster driving. Remove all 'official' restrictions, e.g. traffic lights.
3 Bring in professional racecar drivers to train drivers to go faster. Attributes: Training. Professionalism.	3 Provide training for all drivers on the importance of safe driving, perhaps link to insurance. Anybody caught speeding has to take a course of safe driving before licence is restored.
4 Anyone caught going over the speed limit will be executed Attributes: Meaningful punishment.	4 Understand what punishment would have a real deterrent effect on drivers. Perhaps painting cars pink or add a 'shame' sign saying that they have put people in danger.
5 Make it illegal for men to drive in the city. Attributes: Politically incorrect. Extreme measures.	5 Identify those likely to speed through various types of profiling research (beyond gender), and use that to target deterrent initiatives

side note: in 2008, the town of Bohmte in Germany removed all traffic lights and accident rates declined.

Q My group is generally sceptical about such exercises. How can I convince them it is worth trying?

A Our advice is: do not try too hard to convince! Just do it!

Often, people who are reluctant to give techniques like this a chance have a self-image of a no-nonsense, rationalist, pragmatic 'do-er'. They cannot see a point in these uncertain diversions.

In the face of such in-built scepticism, you might say that we should not spend time debating the pros and cons, which itself might easily take a half hour. Let's do an experiment. Let's do the exercise in the half hour and see what the result is. All you ask for is a half-hour's goodwill.

The follow-on discussion will be much more enlightening, and evidence-based.

By the way, we have done this exercise often and we have rarely seen the half-hour wasted, so long as the participants took part in good faith.

Q How bad is 'bad'? There seems to be no limit!

A The technique requires that you make your bad idea, well, bad!

Of course, it is not acceptable and there is no need to make them cruel or offensive. However, do not be overly polite, as this is constraining. If you think your group may be especially sensitive, you could remind them that everyone is straying into unknown territory with deliberate speed. Any offences or shocks are not deliberate, and will quickly be superseded by the inverted 'good idea'. This is not a competition to cause offence or get a laugh; this is a means of getting to an interesting 'place' that challenges deeply held assumptions, even axioms. Some will be dead ends, but often they can be the source of inspiration for something with genuine potential.

METAPHOR AND ANALOGY   method   1/4

"Metaphor is probably the most fertile power possessed by man", wrote Spanish philosopher and essayist José Ortega y Gasset.

This quote aptly introduces one of our favourite ideation methods. Metaphor and analogy offer access to powerful ideas in innovation by drawing on similarities in attributes, relationships and structures from other domains and applying them back to the innovator's present challenge.

Innovators, like other designers, artists, entrepreneurs, authors and leaders, have always used metaphor and analogy in their work to create and communicate original ideas. It requires identifying important attributes of a problem, solution, system or context under study and recognising that all or some of these attributes are present in another, metaphorical context. Study of this metaphor often brings fresh insights. For example, a device to help drivers know their position on the road at night-time was developed with the help of the metaphor of 'cat's eyes'. As well, Ford's early high-volume production line for cars used the metaphor of systems for moving meat in abattoirs, which had been around for millennia. In another context, Apple's Steve Jobs famously described a computer as a tool that is 'a bicycle for our minds'.

What about the comparison between a new medical clinic and McDonald's? Entrepreneur Rick Krieger found himself waiting for hours in A & E, only to discover he had the easily treated condition, strep throat. He thought there must be a better way to do healthcare without wasting patients' hours in queues and wasting the skills and time of trained doctors for simple ailments. He pondered which other domain shares similar attributes, and how might it deal with them? He thought of McDonald's, which deals with very busy periods, has low-paid workers but is fast, affordable and consistent. Drawing inspiration from McDonald's, **MinuteClinic** would offer a limited menu. Dealing only with five common illnesses, it would be cash only, it would hire lower paid nurse practitioners and it would be easily accessible in cities. MinuteClinic went on to be sold to CVS Health for an undisclosed but, we guess, healthy sum.

Step by step

1. **Set-up**
   Identify the challenge clearly. Teams of 4–6 work well. Divide larger groups into teams of 4–6.
2. **Identify attributes**
   Identify undesirable or problematic attributes of the current problem situation or context of your challenge. Avoid simplistic or generalised attributes like 'difficult to use' and 'high cost'. Instead, push for attributes that are distinct, specific to your problem, and at a profound or systemic level. Identify three, at least, but more if possible. [10 minutes]
3. **Identify similar problems in distant domains**
   Task the teams to find three problem scenarios from distinct, distant domains, which have multiple attributes in common with the ones you

identified above. This task requires intense creative effort. It needs practice and benefits from vibrant, imaginative team discussion. Sometimes, a
team may land on an intuitively excellent metaphor quickly. However, if
imagination stalls, the facilitator may nudge the groups to search for inspiration in ripe domains such as nature, military, sport, or familiar services
like ambulance, pizza delivery, church services, etc. [20 minutes]

4. **Confirm the metaphor's common attributes**
   Review the three metaphorical problem scenarios identified, and re-
   articulate the 'problem' attributes of each that are in common with your
   core challenge. [10 minutes]

5. **Reverse the metaphor to stimulate ideas**
   How are the problem attributes handled in the distant domains? Taking each in turn, identify how the problem is handled. Now, reverse the
   metaphor and consider what this might suggest for your own challenge.
   What possible solution ideas, features, configurations or qualities does
   it bring to mind? Keep adding distinct new ideas until you run out of
   steam. [20–30 minutes]

METAPHOR AND ANALOGY template 2/4

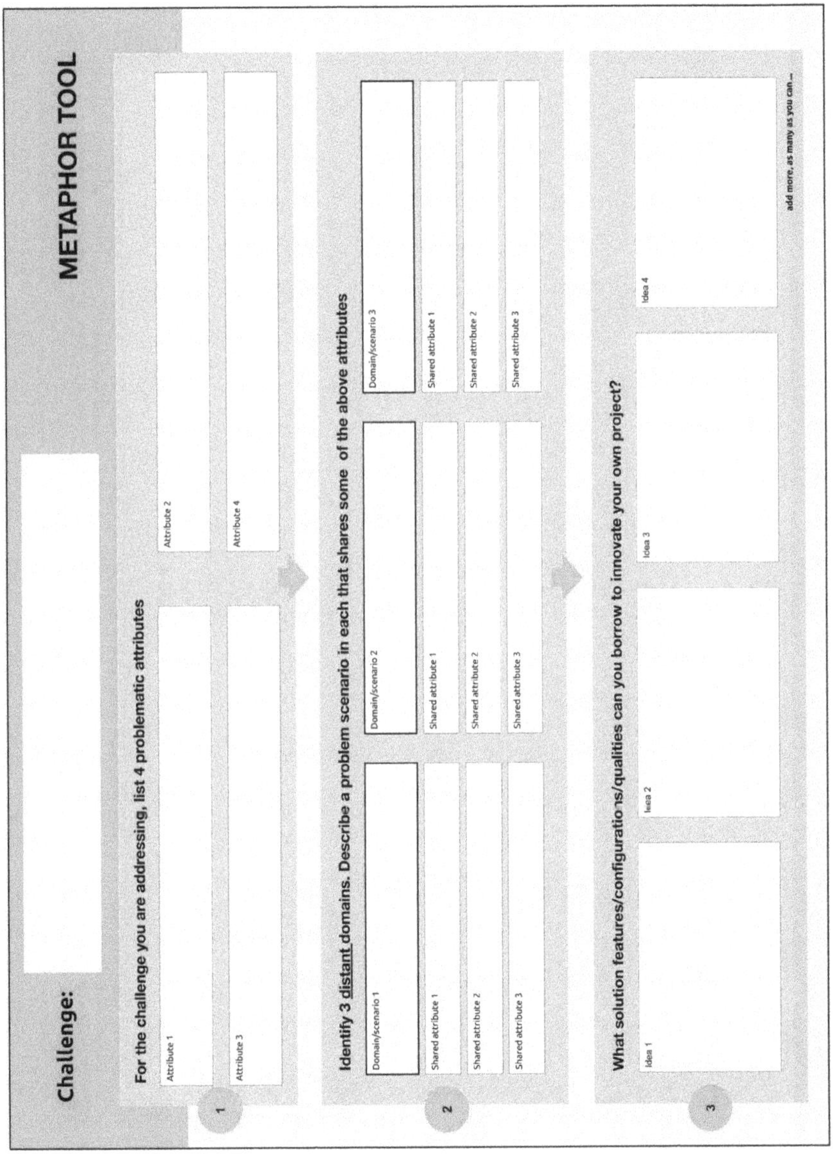

FIGURE 5.14 Metaphor method template.

METAPHOR TOOL

Challenge:

For the challenge you are addressing, list 4 problematic attributes

1

Attribute 1

Attribute 2

Attribute 3

Attribute 4

Identify 3 distant domains. Describe a problem scenario in each that shares some of the above attributes

2

Domain/scenario 1

Domain/scenario 2

Domain/scenario 3

Shared attribute 1

Shared attribute 1

Shared attribute 1

Shared attribute 2

Shared attribute 2

Shared attribute 2

Shared attribute 3

Shared attribute 3

Shared attribute 3

What solution features/configurations/qualities can you borrow to innovate your own project?

3

Idea 1

Idea 2

Idea 3

Idea 4

add more, as many as you can...

FIGURE 5.15 Metaphor method example.

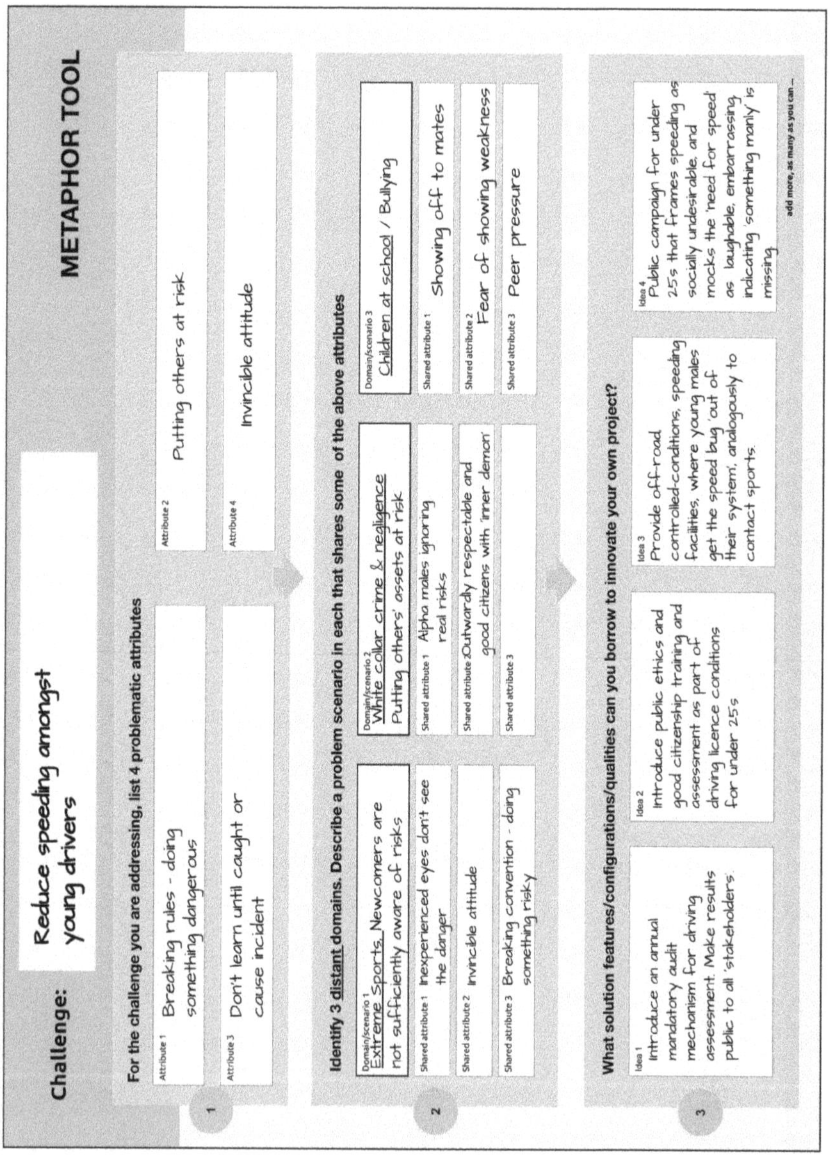

METAPHOR TOOL

Challenge: Reduce speeding amongst young drivers

For the challenge you are addressing, list 4 problematic attributes

Attribute 1 Breaking rules - doing something dangerous

Attribute 2 Putting others at risk

Attribute 3 Don't learn until caught or cause incident

Attribute 4 Invincible attitude

Identify 3 distant domains. Describe a problem scenario in each that shares some of the above attributes

Domain/scenario 1 Extreme Sports. Newcomers are not sufficiently aware of risks

Shared attribute 1 Inexperienced eyes don't see the danger

Shared attribute 2 Invincible attitude

Shared attribute 3 Breaking convention - doing something risky

Domain/scenario 2 White collar crime & negligence Putting others' assets at risk

Shared attribute 1 Alpha males ignoring real risks

Shared attribute 2 Outwardly respectable and good citizens with 'inner demon'

Shared attribute 3

Domain/scenario 3 Children at school / Bullying

Shared attribute 1 Showing off to mates

Shared attribute 2 Fear of showing weakness

Shared attribute 3 Peer pressure

What solution features/configurations/qualities can you borrow to innovate your own project?

Idea 1 Introduce an annual mandatory audit mechanism for driving assessment. Make results public to all 'stakeholders'

Idea 2 Introduce public ethics and good citizenship training and assessment as part of driving licence conditions for under 25s

Idea 3 Provide off-road controlled-conditions, speeding facilities, where young males get the speed bug 'out of their system', analogously to contact sports

Idea 4 Public campaign for under 25's that frames speeding as socially undesirable, and mocks the 'need for speed' as laughable, embarrassing, indicating something manly is missing

add more, as many as you can ...

METAPHOR AND ANALOGY faqs 4/4

Q What are the pros and cons of the metaphor method?

A Metaphors are very powerful because they invoke a natural way of thinking. We all understand a situation by seeing patterns of its elements' configuration and relationships. When we identify partial overlap between the pattern of one situation and another (metaphorical) one, it allows our mind to play with the extended pattern so as to discover further relevance. The great advantage of the metaphor is the immediate opening up of a new, extended pattern of relationships for consideration in your own challenge.

Note, the extended pattern must be for consideration only! You should always be very careful that you do not stretch the metaphor too far so that it would be invalid.

Q You seem to interchange 'metaphor' and 'analogy' when you are discussing the technique. Why?

A You will find both terms are used by different authors. Analogy identifies explicit similarities between two situations, while metaphor assumes equivalence without expressly pointing out the basis for it. The difference is subtle and, though we see metaphor as a more powerful and inspiring tool, we think the difference is not significant enough to agonise over for our present purpose, and we are happy to use them interchangeably.

Q Are there other ways to draw inspiration from another domain?

A Yes, of course.

An alternative approach that is similar to the method described in this section is to initially identify attributes of an *ideal solution* to your challenge. What would a great solution look like, in outline? First, search for a distant domain example where such a system, service or product exists. Then bring back the useful observations, as to how it was achieved, to your own domain. For example, a team from Dell wanted to achieve very fast response and support delivery times in cases of customer emergencies. So, they drew on the metaphor of an ambulance service to create a network of engineers ready and equipped to deal with such emergencies.

Another example is 'biomimicry', which involves direct imitation of systems, elements and behaviours of nature for the purpose of solving complex problems.

Recently, the COVID-19 pandemic has been likened to a war, and many commentators suggest that the war-readiness elements found in national and international military systems should be applied to pandemic-readiness by national and international public health systems.

Converge on three or four coherent concepts that you and the team will further explore, test and develop.

The output of a successful and rigorous idea generation phase should be a host of plausible, or at least conceivable, ideas or idea-fragments, each of which might form part of a solution. Ideally, for what follows, the ideas will be captured on Post-it notes, although other movable card systems will also work. From a wide range of plausibility across all idea fragments, it is likely there are a few that might be gateways to an exciting new innovative outcome.

When faced with hundreds of disparate ideas, it can be daunting to attempt to decipher patterns from amidst the plenty. Making sense of the idea cloud, identifying patterns and themes, forming early concepts and prioritising those that deserve further attention, is the process we call synthesis. Conceptually, the process of synthesising idea fragments into concepts is similar to synthesising research data into insights. Physically, it is analogous to building Lego structures from the elemental Lego bricks.

A key objective of concept synthesis is to create solution choice, not just one 'solution'. Multiple concepts will be explored with the user to identify an optimum final solution based on iteration of one or a combination. Sometimes, it is difficult to get your attention away from an 'anchor' concept that was the favourite from the start. Our M3 technique helps avoid this by encouraging focus on different innovation ambitions, as represented by Mild, Moderate and Mad concepts. See further detail earlier in this chapter.

Mild: Provides incremental improvement to the users' dilemma, with 3–9 months' effort to deployment.

Moderate: Provides substantial improvement to the customer and significant differentiation to you as supplier, with 12–24 months' effort.

Mad: Provides almost perfect satisfaction of the Reframed Vision without compromise. Has compelling desirability, with feasibility constrained only by ten years' well-resourced effort.

Step by step

1. **Prepare**
   Prepare the room and the team for a 2–3-hour closed session. Be sure the Reframed Vision Statement is written clearly on the wall as a continuing focus. Synthesising ideas into concepts is analogous to the synthesis of research data into insights, described in chapter 4, Reframe.
2. **Display ideas on the wall**
   Gather all idea fragments (Post-its) and place them on the wall, one at a time. Speak out loud what is on the Post-it, in order to share with the team, but do not discuss at length – it is only an idea! Place similarly themed ideas physically close to each other, when this is obvious, but don't worry over this for long. Keep moving. Remove obvious duplicates.

3. **Arrange into themed clusters**

   Through organic discussion and debate, have the team re-arrange the post-its into themed clusters. A theme represents a common emphasis on a particular feature, user need, angle of attack, or outcome. Allow for constructive debate and argument, but beware that a team leader may eventually have to make a call so as to maintain progress. Name each cluster.

4. **Decide on anchors**

   Choose an 'anchor' cluster for each of your proposed alternative concepts, one at a time. These may be 'mild', 'moderate' and 'mad' concepts, or other compelling alternatives. Choose the anchor using the team's judgement that it will provide maximum impact towards the Reframed Vision, and may be realised within the indicated timeframe.

5. **Build the concepts**

   For each anchor, build out and enrich each concept by adding new ideas and reconfiguring, paying particular attention to its desirability, viability, feasibility and strategic suitability characteristics, as far as can be judged in the moment. Use sketches to represent and refine your concepts.

CONCEPT SYNTHESIS template

FIGURE 5.16   Concept synthesis process using Lego analogy.

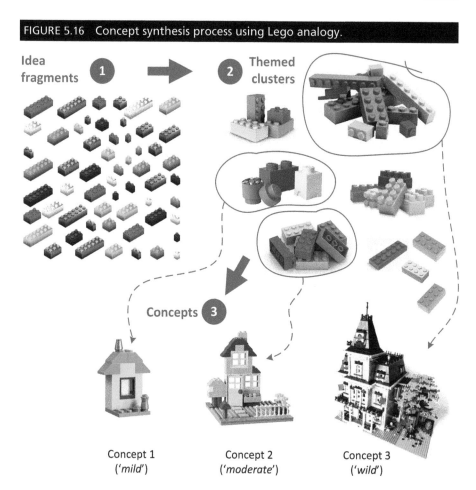

Idea fragments ①

Themed clusters ②

Concepts ③

Concept 1
(*'mild'*)

Concept 2
(*'moderate'*)

Concept 3
(*'wild'*)

CONCEPT SYNTHESIS example 3/4

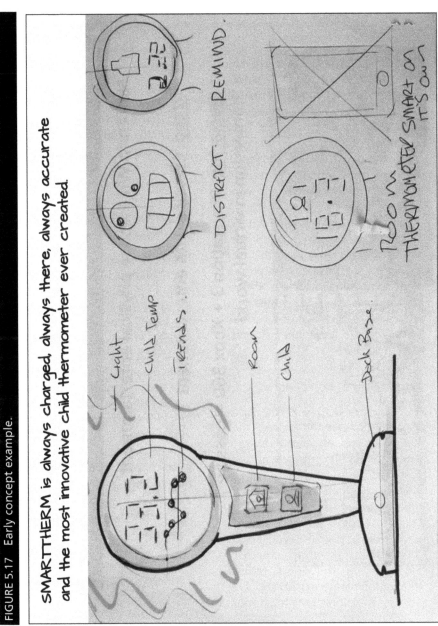

FIGURE 5.17 Early concept example.

SMARTTHERM is always charged, always there, always accurate and the most innovative child thermometer ever created.

Q Are we not better off concentrating all our efforts on one good concept rather than three?

A Concepts, at this stage, are the direct output of your ideation. Though based on solid research and reframing, still they are the fruits of imaginative leaps. They have only limited value until you prove them with users (Validation). So, you must stretch yourself to provide a range of options to your user for feedback. When you bring your proposal to your user, you do not want just a binary response, yes or no. You must allow the user a real choice through comparing and contrasting multiple different options. Also, this gives you a greater, richer base of feedback with which to iterate the next version of concept.

Q What is the point of a Mad concept?

A You might reasonably guess that ten years and an almost infinite budget is highly impractical. But, there is a useful purpose.
Your imagination is framed and bounded by your particular knowledge set and experience. This is useful if you are in the domain of your expertise. However, by definition, radical innovation breaks existing boundaries and explores new territories, where there can be no experts. Our advice is to ignore, for a while, traditional technical or operational constraints and to explore without these boundaries, paying attention only to the user desirability requirements. If you find a concept that is truly compelling, there is usually a clever way to sidestep a technical or other barrier.
In other words, do not limit your imagination too soon.

Q Please explain the Lego analogy some more.

A Lego play is about concept building. When a child, or adult, plays with Lego, she is trying to give physical form to a concept within her head, a figment of her imagination. This is exactly what you are trying to do with your innovation concepts here.
The idea fragments from ideation are represented by the building blocks. Some are simple, some are more complex. Individually or in groups, some are chosen by your team as platforms or anchors for the concept you wish to build. You build a concept by choosing a base (idea) and adding others to strengthen and enrich it.
It is a very powerful analogy that will provide useful guidance for you and your team through this concept building phase.

This is a clear visual presentation of the most important elements of a concept.

Early concepts in an innovation project determine the future direction of the project. All further work and outcomes flow from the early concepts as they are whittled towards determining the final solution offering. A concept must hold promise that it can satisfy the combination of criteria to be desirable, viable, feasible and strategically suitable. At this early stage, most emphasis has been on the desirability dimension, although 'sanity check' background attention must be given to the others. See the next method section of this chapter.

You will have multiple early concepts that you wish to bring to the user for feedback. They may be complex, based on detailed research, insight generation and reframing. They may have a number of nuances and subtle acknowledgements of insights. The concept board is a means to cut through this complexity. It gives an easy-to-absorb overview of the key points and justifications, through a highly visual presentation.

Note, the words 'concept board' have various understandings in different organisations. At this early concept stage, our concepts are firmly oriented towards the user, to facilitate getting feedback on what works and what does not work. We call this *Concept Board (A)*.

At a later stage, after concepts have been tested, validated, refined and reduced to a final selection, an extended version of the concept board is used to provide internal (to the firm) explanation and investment justification. We call this *Concept Board (B)* (see chapter 7, Execute).

Concept boards are single page, in a size and style that suits the way you intend to use them. When bringing our proposition to users we mostly like to use A3 size on 12 mm foam backing, which is low cost and quick to produce. Other formats may suit different situations better.

Step by step

1. **Who will create it?**
   The concept board is not a production-quality item, but it must be clear. Clarity involves removing clutter of detail and visually highlighting the core value proposition and key benefits to the user. We suggest a template on the next page, but there are many alternative formats that fit the bill.

   It is best produced by a designated 'artist' following direction from a team discussion. The artist, ideally, will have an artist's or designer's keen awareness of what is core and the ability to represent it clearly. There is no fine artistic skill or CAD tool required. It may be produced using Microsoft PowerPoint or Google Slides, or equivalent. The concept board, once created, will be agreed with the team.

2. **Concept briefing meeting**
   Convene a team meeting, including the designated artist. It may be immediately after concept synthesis, but it can be useful to allow thoughts to mature overnight so that the team comes back to the task with fresher minds.

Discuss the concept with the team. What is the concept's essential value proposition to the user? What are the key benefits to the user and what research insights does the concept address? Keep pushing until these are specific and clear. Agree a meaningful and evocative name to make it feel real and to facilitate comparing and contrasting it in discussion. The artist allows the meeting to close when she is fully clear on all points.

3. **Concept review meeting**

For a simpler concept or a particularly skilful designer, the concept board may be created in real time. More usually, the artist will retire and come back to a later review meeting with her proposal. Usually this captures most of what the team intended. However, she or the team may have had some further thoughts on additions or refinements, and these may be incorporated rapidly before signing off.

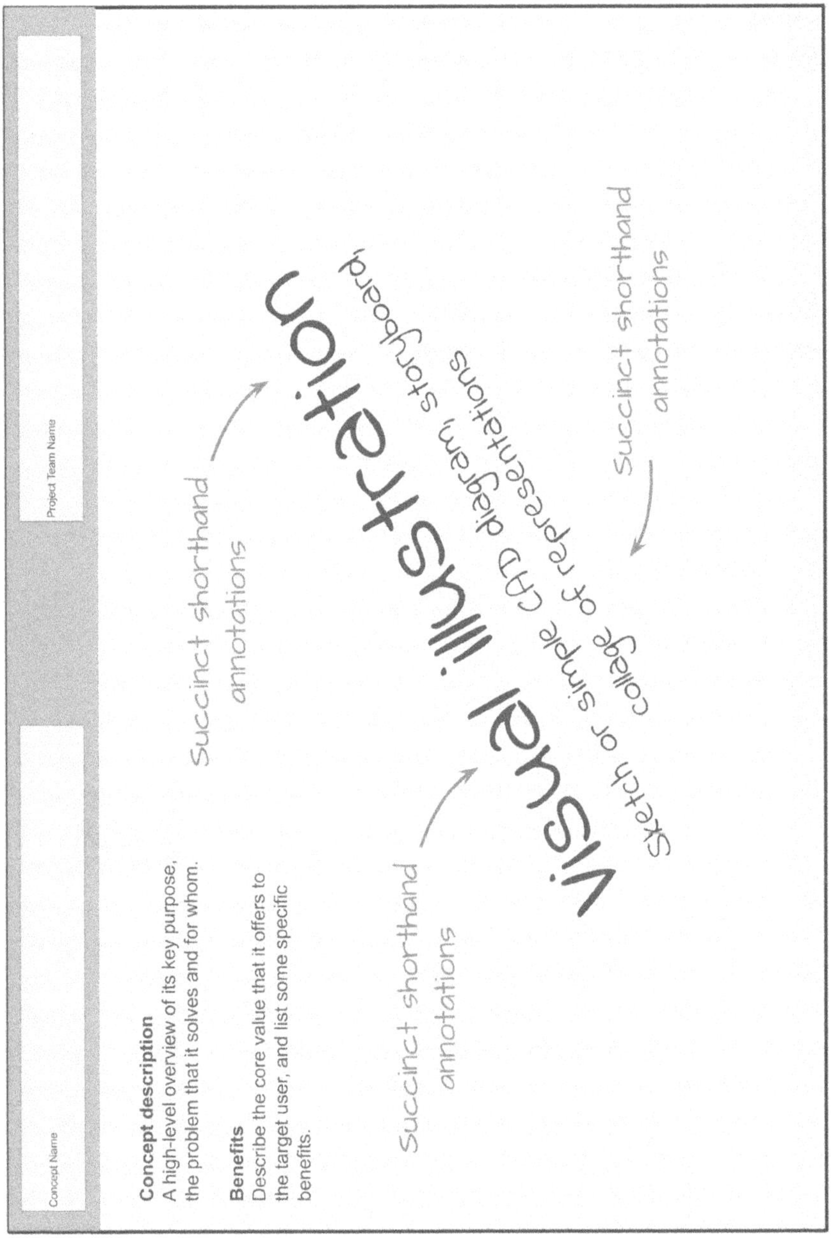

FIGURE 5.18   Concept Board (A) template.

Concept Name

Project Team Name

Concept description
A high-level overview of its key purpose,
the problem that it solves and for whom.

Benefits
Describe the core value that it offers to
the target user, and list some specific
benefits.

Succinct shorthand
annotations

Succinct shorthand
annotations

Succinct shorthand
annotations

Succinct shorthand
annotations

visual illustration

Sketch or simple CAD diagram, storyboard,
collage or representations

FIGURE 5.19 Concept Board (A) example.

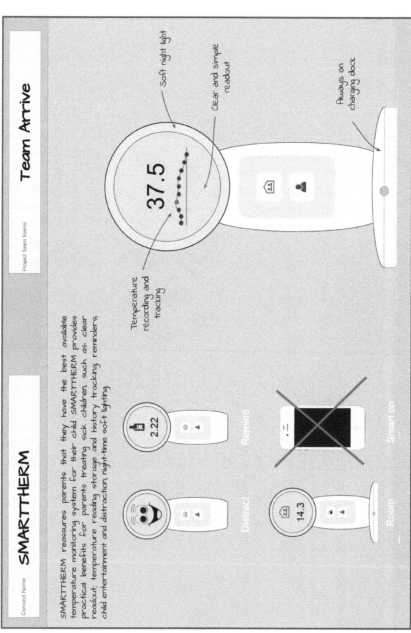

SMARTTHERM

Concept Name

Project Team Name

Team Arrive

SMARTTHERM reassures parents that they have the best available temperature monitoring system for their child. SMARTTHERM provides practical benefits for parents treating sick children such as: clear readout, temperature reading storage and history tracking reminders; child entertainment and distraction, night-time soft lighting

Soft night light

Clear and simple readout

Always on charging dock

37.5

Temperature recording and tracking

2.22

Distract

Remind

14.3

Room

Smart on

Q Our staff are scientists and engineers. They are well able to create technical drawings, but have no practice at drawing for laypeople. What should we do?

A Remember a few points:

- These are not meant to be finished production drawings or realistic artist's impression pictures.
- Low fidelity is better than near realism. If it appears almost real, the viewer is likely to expect a fully real experience and is disappointed when this is not achieved. With low fidelity (like 'impressionism' painting), when you stress the essential features the viewer interpolates the remainder to complete a satisfactory mental picture.
- You are not trying to deceive the user that you have something definite and complete to offer. Your purpose is to offer an image or impression and to ask the user: "Am I on the right track", or "How would you suggest we might improve what we are suggesting here?"

With the above in mind, do not be daunted by the skill or production quality required. Probably, one of your team is well able to do it. If not, a skilled artist or designer can perform a very good job with a few hours' attendance at your briefing and review meetings (steps 2 and 3). Make sure you brief them properly, as above.

Q How much detail is enough?

A Less is more. A concept should have one overarching and clear value proposition, supported by other subsidiary features.
Be sure the overarching value proposition is clear and not smothered by other detail. The most common mistake is death by a thousand darts, where the main value proposition cannot be identified amongst the clutter.
Think of a core value proposition, a skeleton shape to support it, and some (few) additional features to put flesh on the bones.

Q How long should it take to create a concept board?

A Please do not treat the concept board (i.e. the physical item) as something precious. It is only a tool to get feedback and advice from the user. So, do not let the physical production delay you.
In this method introduction, three pages back, we described two meetings – one for discussion and briefing, and the second for sign-off. These should be less than 1 hour and a half-hour, respectively.

This is a regularly referenced health check to ensure there are no insuperable barriers – deal breakers – to a concept's eventual successful conclusion.

In this book, we repeatedly emphasise that a good project must be desirable, feasible, viable and strategically suitable. We prioritise desirability, by emphasis and sequencing, because we know that insightful understanding of the user's desired outcomes is a powerful driver of a successful project. Of course, you must never fully ignore the other factors. However, it can be easy to do so in the midst of the team's excitement and enthusiasm around delighting your user.

Experienced innovators know this intuitively, and they automatically, constantly, take tacit account of all possible factors. Less experienced innovators and teams sometimes have more difficulty. So, we recommend regular health checks to maintain team alertness and sensitivity.

Note, we do not wish at an early concept stage to stifle flows of creativity and enthusiasm for solutions that delight the user. This would surely happen if all FVS potential issues were to be regularly investigated in detail. On the other hand, however, it would be folly to pursue a concept for too long where it was clearly technologically infeasible or otherwise unachievable. What is required at this stage is a coarse evaluation – effectively 'go/no-go' – the key result of which is a considered opinion that there is, as yet, no insuperable barrier foreseen. This check might be done informally on a daily or weekly basis, but in any event we recommend at least monthly, by more formal team meeting.

Our index provides a range of key considerations under the three headings (FVS) and suggests a score based on the evaluation assigned to each. For many teams, the go/no-go decision will not directly follow from the resulting numerical index value, but the result should certainly give pause for thought, which is its main purpose.

Step by step

1. **Team meeting**
   When an early concept is considered worthy of further development or investigation, set aside one hour for the team to evaluate its FVS Sanity Index. Repeat this regularly during concept development – at least once per month, or more frequently, depending on the project scale or complexity.

2. **Discuss the nine questions in turn**
   Remind the team of the concept under consideration and of the purpose of the meeting. Display the concept board in large format for all to see and focus on, as well as any other relevant artefacts or notes.

   The FVS Sanity Index contains nine questions, three for each category. The template is on the next page. Review the questions, giving detailed attention and discussion to each. If there are other important questions for your project, introduce them at the appropriate time.

Note, you will not be able to answer all questions exactly! And, some people will be reluctant to engage in speculation on some questions. Nevertheless, the team leader must push for an agreed conclusion. While it may not be possible to say for sure what a correct answer is, in many instances it is possible to give a range or a threshold with a high degree of confidence. For example, you may not yet know exactly all development activities to bring a new concept to deployment, but you may still be able to judge with confidence that it appears comparable in scale to another development done last year that was brought to market in eight months. In this case, as a threshold, you might confidently say that the time to deployment will certainly be under 12 months.

3. **Review and decide**
   Review each category score and overall index score. Are there any deal breakers to indicate the concept must be re-thought? This might arise from individual low scores or overall weak score. If some aspect remains unknown or highly uncertain, then this also merits immediate investigation before going further.

FVS SANITY INDEX template

FIGURE 5.20   FVS Sanity Index template.

FVS Sanity Index
(sanity checking Feasibility-Viability-Suitability factors of early stage concepts)
Score < 15 in any category = serious problem!
Score >20 in all categories = looking good …… if the user wants it!

Concept name:

Feasibility

Are the likely implementation technologies …		What is the level of execution risk?		What is the time from development start to deployment readiness?	
New to the world?	1	Very high	1	> 24 months	1
Stretch for the firm?	3	High	3	<18 months	3
New to the firm?	6	Moderate	6	< 12 months	6
Routine for the firm?	10	Low	10	weeks	10

Comments

Feasibility score (/30)

Viability

How scalable is the concept?		What is the margin possibility?		Is the required investment …	
Not much – very specific	1	Tight	1	Special?	1
Moderate	3	Moderate	3	Large?	3
Good potential	6	High	6	Routine?	6
Exceptional	10	Exceptional	10	Minimal?	10

Comments

Viability score (/30)

Strategic Suitability

How well is this concept aligned with our current mission and strategy?		How great are the organisational changes necessary?		Do we have/will we get enough high-level support in the firm?	
Not clearly aligned	1	New division/company	1	Not clear, not explored	1
Not clear	3	New team	3	Broad team support	3
Aligned, but new activtity	6	Moderate	6	High-level champion	6
Fully, clearly	10	None	10	Clear, strategic, CEO level	10

Comments

Suitability score (/30)

FVS SANITY INDEX   example

FIGURE 5.21   FVS Sanity Index example.

FVS Sanity Index
(sanity checking Feasibility-Viability-Suitability factors of early stage concepts)
Score < 15 in any category = serious problem!
Score >20 in all categories = looking good ...... if the user wants it!

Concept name: *Alpha Vision System*

Feasibility

Are the likely implementation technologies ...		What is the level of execution risk?		What is the time from development start to deployment readiness?	
New to the world?	1	Very high	1	> 24 months	1
Stretch for the firm?	X	High	3	<18 months	5
New to the firm?	6	Moderate	X	< 12 months	X
Routine for the firm?	10	Low	10	weeks	10

Comments
There are some significant new technologies for the firm. But, the scale is limited and we expect we can buy in those essentials that we don't have in-house.

Feasibility score (/30)	15

Viability

How scalable is the concept?		What is the margin possibility?		Is the required investment ...	
Not much – very specific	1	Tight	1	Special?	1
Moderate	3	Moderate	3	Large?	3
Good potential	X	High	X	Routine?	X
Exceptional	10	Exceptional	10	Minimal?	10

Comments
Market indicators are all good. This can be our new flagship!

Viability score (/30)	18

Strategic Suitability

How well is this concept aligned with our current mission and strategy?		How great are the organisational changes necessary?		Do we have/will we get enough high-level support in the firm?	
Not clearly aligned	1	New division/company	1	Not clear, not explored	1
Not clear	3	New team	3	Broad team support	3
Aligned, but new activtity	6	Moderate	X	High-level champion	6
Fully, clearly	X	None	10	Clear, strategic, CEO level	X

Comments
The project is firmly promoted and backed by CEO and Marketing. The strategic plan effectively mandates this or similar project to drive mid term growth.

Suitability score (/30)	26

This FVS Sanity Index shows the thoughts of an Irish firm's project team tasked to develop a new visual inspection system for manufacturing use.

While the Alpha programmable concept represented a new direction for the firm, with new features and capabilities for the market, the overall FVS Sanity Index, happily, also indicates a strong potential for success deployment.

Q When is the right time to do this check?

A We don't like to think of a tightly prescriptive time scale for this. From one perspective you should be doing it every hour, every day. Experienced innovators, like many experts, have an intuitive 'sixth sense' that alerts them on the fly to potential anomalies beyond the immediate issue at hand.

But, even experts must be corralled occasionally to clarify their thoughts and sense-check intuitions, particularly when working as part of a team. So you, as team leader, and all team members must develop this sixth sense that tells you when some concept feature is going outside acceptable bounds. If it appears to do so, it does not mean that the concept should be scrapped. But, it does dictate that the area of concern must be investigated in more detail earlier.

Q What about market size, revenue, business model and other important parameters? What is the point in excluding these?

Where would we stop? For a summary evaluation that does not interrupt the flow of creative concept development, we think our index questions are good thought provokers. But, you should not hesitate to add more that may be important to your specific project.

Q Should we bring in specialist help, for example to help with questions of technology availability or financial investment?

A Yes. It is good to have, for example, a technologist and a financial/cost accountant on the team. Alternatively, if they are not permanently active team members, they may be regular consultants. The same applies to other specialists depending on what may be missing from the mix of your core team.

However, be cautious to ensure that an external consultant is properly briefed about the process and the stage you are at. Many specialists see it as their role to be extra cautious about issues in their domain. Emphasise that the purpose at hand is not to accurately predict an exact future outcome or configuration. The purpose is to establish the probable bounds of the range of outcomes and to judge whether this is likely to lead to a successful deployment. Finessing and implementation details come later.

Obviously, any major anomalies need to be investigated before further investing in this concept.

BIBLIOGRAPHY

Adner, Rod and Kapoor, Rahul (2016). *Right Tech, Wrong Time. Harvard Business Review*, November.

Amabile, Teresa (1998). *How to Kill Creativity. Harvard Business Review*, September.

European Commission (G) (2014). Technology Readiness Levels (TRL). Horizon 2020 – Work Programme 2014–2015 General Annexes, Extract from Part 19 – Commission Decision C(2014)4995.

Gunther McGrath, Rita (2010). *Idea Generation is Seldom the Problem*, viewed 25 May 2020. https://www.ritamcgrath.com/2010/07/idea-generation-is-seldom-the-problem/.

Knapp, Jake (2015). *Sprint – How to Solve Big Problems and Test New Ideas in Just Five Days*. Bantam Press.

CHAPTER **6**

Validate

Prototype regularly, as quickly and minimally as possible.

A product concept describes the core value proposition to the user. The business proposal describes how that concept will be delivered to the users, with other key business parameters. It adds considerations of market viability, technology feasibility and strategic suitability for the firm.

All assumptions that are embedded within the business proposal and that can be identified as success factors for the project must be made explicit and then validated to the fullest extent possible through a process of iterative prototyping, testing and refinement. In this chapter, we provide details of prototyping designed to test *Desirability* assumptions in particular. We use pro-forma *Assertions Worksheets* to draw out the assumptions associated with common parameters of a business proposal, and we use the *Lean Canvas* as a reference for developing the business model.

FIGURE 6.1   Chapter 6 outline.

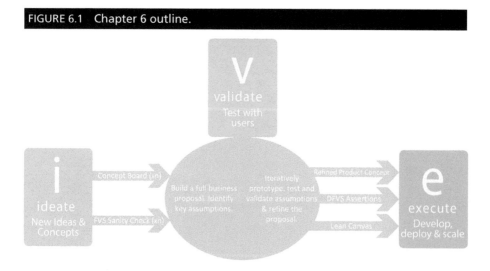

The output from the Validate stage is a relatively robust and risk-reduced (yet still unproven in deployment) representation of the new value offering in the form of *Product Concept* description, *Business Proposal: DVFS Assertions Worksheets*, and an initial *Lean Canvas* business model.

JOANNA'S DILEMMA

It is five months into the project and Joanna's boss, Jack, has not spoken much with her about it, apart from occasional casual enquiries. In his mind, he is happy to let her get on with the work and he has confidence in her experience, expertise and motivation. From her point of view, she has done her regular budget reporting, she has been busy building the team, doing lots of field research and synthesis of findings, again and again, getting ever better understanding of the possibilities. Recently, she has organised a series of ideation workshops with diverse teams of colleagues and guests. It has all been very energising. Time has flown, and she is pleased with the progress. She sees some real potential in the concepts they have shortlisted. Two of them in particular could be game changing. In fact, just last week she could not resist describing one of them to a senior board member whom she met at the coffee machine. He was very impressed!

But, that was a casual conversation too far, and Joanna regrets it today. Jack, the CEO, asked to see her this morning. Jack heard about the new concept and he is delighted with his understanding that the project sounds like it is on track to go to market in month 12, even earlier than he had originally planned, at 15 months. He has had some thoughts about how they might prepare the sales organisation and the channel and wanted to have Joanna's input on that. First, of course, he wanted to hear more detail about the concept and its potential returns, and to arrange for Joanna to present it to the senior management team early next month for detailed discussion.

Joanna is not happy! It is no consolation to know that this is a regular dilemma for innovation project managers like Joanna. This is because there are common misunderstandings of what a *concept* really is. For Jack it is a decisive direction – major decision made and groundwork done; only operational details, like manufacturing planning and a go-to-market strategy, remain to be ironed out.

For Joanna, the concept is still unfinished. It is an ongoing developing platform for deeper exploration and evolution into a full business proposal addressing all four Factors of Innovation, DVFS. The desirability aspects of the innovation have been fairly well evolved, though not complete, and they need to be confirmed with further field testing. The viability, feasibility and strategic suitability aspects of the business proposal need to be explored, enriched and proven. There is no point in indulging an apparently great concept for too long if it is not technically feasible (for your firm), does not have a clearly sustainable and scalable business model and market, and if there is no appetite or strategic capability in your firm to proceed with the project.

Before Joanna can present confidently to the executive board, she needs time to flesh out a full business proposal around her concept, and she must clearly identify the major assumptions to be tested. Once the assumptions (or substitutions) are validated, she can be more confident recommending to the board the investment required to deploy the project.

BUILDING THE BUSINESS PROPOSAL

In this book, we do not treat in detail how to choose the best business models, conduct rigorous market or financial analyses, prepare technology evaluation and roadmaps or other more traditional new product development business exercises. These are well covered in many other excellent books. However, through this chapter we focus in particular on the desirability dimension of the proposal and we give you a framework and a plan for how to organise and prioritise your work in all four dimensions.

Under the ARRIVE process for design innovation, you started with the Audit and Research phases, where you accumulated a good understanding of your user, target market and general environment, all under the headings D-V-F-S. From this, you developed a concept (or two or three) that you have good grounds to think ought to be compellingly desirable for the user.

Building on the above, you should be able to flesh out initial estimates of the elements of a full business proposal. These will be encompassed in answers to questions like these below, or at least sensible guesses that come from your team's recent immersive experience. Of course, all these questions may not be relevant to your project, and there may be some other questions that you should add, for your particular case.

Desirability questions

Who will our target customers be?

What jobs will our solutions help them complete?

What are the current obstacles for our customers to completing these jobs successfully?

How will our customers measure success of the outcome?

What are the customer's main alternatives?

Is the user experience that we offer demonstrably better, simpler or more convenient than the alternatives?

What overwhelming advantage do we have that will compel our customers to switch to our solution?

Viability questions

Who or what is our primary competitor and broader competition profile?

How will the competition respond to our solution?

How will we generate revenue?

How will we acquire new customers; how will we stimulate trial and repeat use?

What will our referral rate for new business be (%)?

What will our repeat business rate be (%)?

How will we achieve growth to X in time Y?

How many paying customers will we have in years 1/2/3?

What will our average revenue per customer be in years 1/2/3 (€/$/£)?
What will our average gross margin be in years 1/2/3 (%)?
Why are we confident our operations organisation has competence and capacity to deploy this project?
What are the project's three-year Investment Amount, Nett Present Value and Internal Rate of Return?

Feasibility questions

What are our significant technical or engineering challenges, and how will we overcome them in time and on budget?
How is our solution technically superior to alternatives?
What competitive advantage does our technology offer?
Why are we confident our team is technically competent to undertake this development job successfully?
What is our development time to deployment?
What is the total amount of funding required for technical development, and from where will we secure this funding?

Strategic suitability questions

Why will our firm's leadership support this project?
Who will be the senior executive champion for the project?
How will this project enhance the firm's brand?
How does the scale and scope of this project fit our organisation and its mission?
How will it be resourced?
What gives us confidence this project will be prioritised sufficiently to attract resources for development?

We provide guidance, in the methods section later in this chapter, on how to build a business proposal using *Business Proposal: DVFS Assertions Worksheets* as a template. *The DVFS Assertions Worksheets* capture, summarise and communicate your answers to the questions above in the form of assertions. To use the worksheets, ask yourself what must be true in order for the project to be a success. (In the FAQ section for this method, we explain why we sometimes use the word *assertion* rather than *assumption*.)

The above assertions encompass many explicit and implicit assumptions across all DVFS dimensions. They are the key premises for success of your innovation project. You will be keenly aware that some of the assertions are well founded, have plentiful backup research, a clear rationale, and may be in no doubt. Others vary in plausibility and probability up to pure, un-evidenced speculation. It is only a summary of each that you need to capture in the worksheets. At this stage, completing a first comprehensive draft of a full business proposal is your objective, so you ride over the many inevitable uncertainties with best estimate assumptions.

Of course, in addition to those above, you should expect to add some of your own assertions, which are specific to your project and exclude those that you consider not applicable. In particular, however, be careful not to avoid an important assertion just because it may be difficult to contemplate or to decide

FIGURE 6.2   DVFS worksheets (see methods section, later in this chapter).

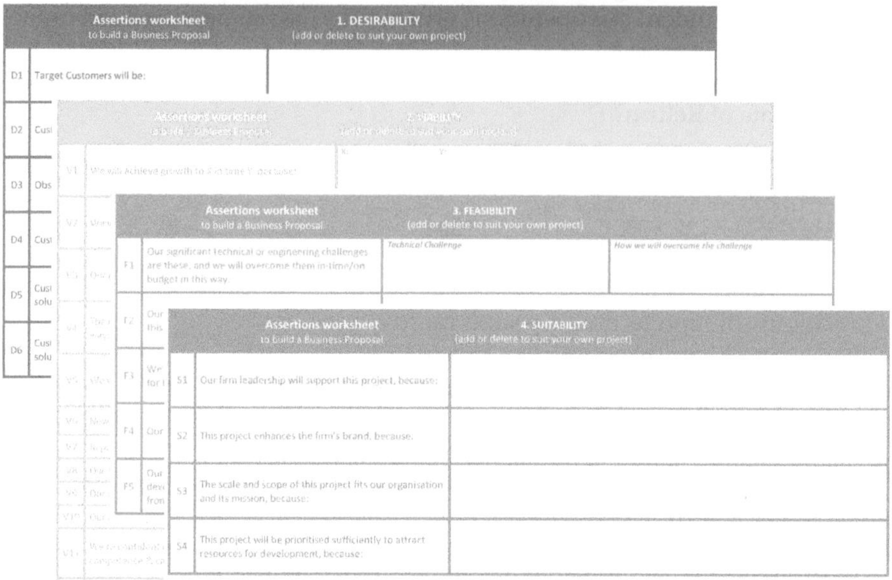

exactly how to represent it! If you do not carefully consider and formulate your position now as fully as possible, it may well default to something less desirable later.

You could think that the research, analysis and answering of these questions is work for experts over many months. This might be true if you were looking for fine precision answers, if such would be ever possible! However, at this stage, your considered estimates are much better than nothing, and there is no one who has studied the situation and possibilities more than you and your team.

The worksheet assertions provide a useful summary record and focus for identifying the most important assumptions. When you and your team are in workshop mode, and seeking to filter and prioritise assumptions, you will transfer these, in even shorter form, to Post-it notes.

VALIDATING THE BUSINESS PROPOSAL

The business proposal is the result of a creative process. Despite the input data being well founded, the creative leaps to arrive at a final concept and an extended business proposal raise new uncertainties. When the premises of success are unproven or un-evidenced assumptions they must be tested, so as to de-risk the proposal. Perhaps the business proposal will need to be amended, or more fundamentally reconstructed, before it can graduate to be considered for sign-off as an approved business initiative.

It is surprising how many people get into difficulty trying to bridge the gap between the more abstract world of new ideas and planning estimates on the one hand, and the practical, business imperative of project sign-off and realising promised return on investment on the other. The *Validation* set of activities is their missing link. The key issues for validation planning are as follows:

- What are priority areas and topics for validation?
- How soon, or in what sequence, may each of them be evaluated and validated?
- Can you rely on existing data or do you need to seek out new data?
- What is the quickest, cheapest, most reliable way to find this new data?

At first glance, the approach may appear straightforward; however, uncovering the best means to find useful answers is often difficult. Validating is mostly a convergent process, but there is still scope for lots of ingenuity to plot a best course for surest de-risking with minimum expenditure.

Remember that, ultimately, it is only in deployment after launch that you get the final proof of your proposal's validity. The proof of the pudding is in the eating! Even at initial launch (Execute phase, see next chapter), you can be sure that your proposal will be sub-optimal in some ways, maybe many ways, but by that stage none of them will be fatal, hopefully. This is the domain of Agile product development and Lean Startup, where new products and businesses are launched but are iterated rapidly in response to experience with real users in the field. We will discuss this further in the next chapter, Execute. For now, in the Validate phase, your task is to increase the prospects for success and reduce the risk involved in progressing your proposal to deployment.

Note that for some projects and concepts, especially simple products, local services or other products where the required pre-launch investment and preparation are low, the validation and early launch (Execute) activities may overlap and merge to form a continuous segue. In cases where a product or business model is more complex and investment-heavy, it is advisable to de-risk through distinct validation testing as much as possible before launch.

SURFACING ASSUMPTIONS

An assumption is a belief that is accepted as true while lacking proof. Experienced innovators know that many of the assumptions embedded in the business proposal at this stage have a tenuous nature, yet they are fundamental premises for success. Still, few practitioners are controlled enough to resist acting as if most assumptions were true, either scientifically proven or axiomatically to be accepted. As the anxiety of broken deadlines and impatience to get to market takes over, too many projects go from ideation to deployment without the crucial validation activities required to detect fatal flaws or to refine a distinctive and compelling advantage.

The crucial task is to isolate the important assumptions and subject them to rigorous testing. After prioritising them according to their importance, you must try to find confirming, or confounding, evidence for each assumption, as early as possible.

We recommend two methods to bring all key assumptions to the surface, ready for the glare of critical scrutiny. Both methods should be used, thereby ensuring that a comprehensive set of assumptions is obtained.

(A) Assertions: "The necessary conditions for success of the project are …"

We have discussed this in the previous section and we describe the method in detail in the later part of this chapter. For surfacing an assumption, it is helpful to bring the associated idea more clearly to the surface by phrasing in the form of an assertion, i.e. make it explicit and claim it is true. Thus, your team has a well-defined target to aim its scrutiny at so as to probe and adjudicate the level of uncertainty and the associated risk.

The outcome of this exercise is a set of completed worksheets, covering the Desirability-Viability-Feasibility-Suitability dimensions that constitute the essence of your business proposal.

(B) Pre-mortem: "The project failed because …"

Have you forgotten anything? Is there anything you have not considered that could cause the project to fail? The worksheets above are a good way to ensure that standard or common factors are considered. An innovation project, however, has plenty of unusual, non-standard and unpredictable circumstances surrounding it, some of which have potential to derail the project from the path of success.

A pre-mortem (Klein, 2007) is a technique that uses *prospective hindsight*, which increases by 30 per cent an individual's ability to correctly identify reasons for future outcomes. The pre-mortem asks an innovation team and other knowledgeable stakeholders to imagine themselves in the future and to posit that a project has failed, either marginally or drastically. The group then brainstorms to come up with reasons as to why it might have failed. Participants are encouraged to include any reason whatsoever, no matter how eccentric, politically incorrect or contrary to perceived wisdom.

In a second round, the actions or means to mitigate each failure mode are considered and a solution is asserted for each. These become additional conditions for success of the project, which may be added to those already identified under the 'assertions' method above. In a workshop, these new assertions are captured in Post-it notes.

The pre-mortem encourages participants to change frame into a failed future and use their ingenuity actively to seek out and highlight areas of greatest weakness. The undesirable alternative, which is commonplace, is that team members may instead apply that ingenuity counterproductively to justify to themselves and others why the possible weaknesses may have been covered in work already done. The latter is often the case when a team is heavily invested in the project thus far and deadlines are tight. The change in perspective with a pre-mortem can unearth some surprising and valuable results.

A fuller description of the method is given later in the chapter.

MAPPING ASSUMPTIONS

When all conditions necessary for the success of the project have been surfaced in the form of assertions, using the two methods above, the core task is to sort them according to their level of uncertainty and importance to the project's success. A successful project is one that is brought to deployment within predicted time and budget, and performs in the market and for the firm as described in the proposal.

Let us understand the range of 'importance' for this purpose. Looking at it one way, everything is important or it should not be there, but it is more helpful to understand importance relative to available alternatives and timing. For example, in the case of the fabled horse that lost the big prize in the Gold Cup race because he lost his shoe in the last lap due to a missing nail in the shoe, that nail was of crucial importance at the end. But another nail at the right time – before the race started – could easily have avoided all mishaps. So, in this example, a missing nail would be assigned a low importance earlier in the project because there are plenty of alternatives at the right time. On the other hand, a proper checklist and quality control process even at an early time should be assigned a high importance.

An example scale of importance is given in Figure 6.3. A high importance score means a factor is critical for success and there is no easy alternative that does not seriously impact the budget, time to deployment or the return on investment after deployment.

The 'uncertainty' scale provides another perspective on the critical success factors of the assertions. How confident are you that your assertion is true? It is important to stay grounded in this discussion and not to get caught in fruitless or impractical philosophical argument, which we have seen occur at times. The philosophical idea of *solipsism* holds that nothing can be shown to exist or be true outside of the human mind, but that is not immediately useful! What we are seeking here is a perspective of reasonableness: what can

FIGURE 6.3 Scaling uncertainty and importance dimensions of assumptions.

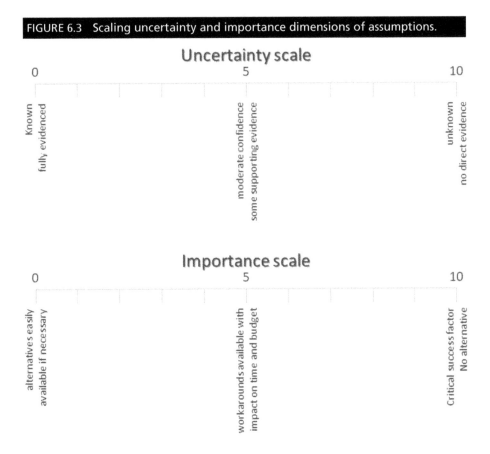

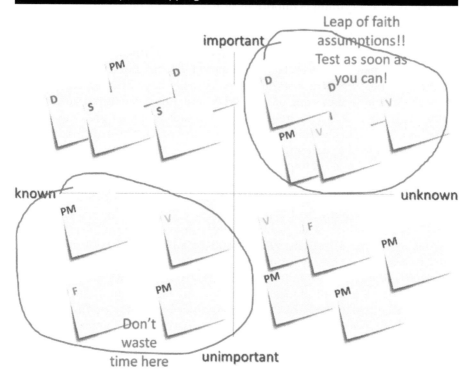

FIGURE 6.4 Assumption Mapping.

we reasonably rely on as fact? Everything else we consider an assumption with incomplete supporting basis. An uncertainty scale is shown below. At one end, there is no uncertainty when something has been fully evidenced or proven. At the other end, something is highly uncertain if it is pure speculation without any supporting evidence or justification. In between the two extremes there will be different levels of supporting evidence leading to varying degrees of uncertainty

Some organisations prefer to sort by applying a numerical score to every assertion for each of the parameters 'uncertainty' and 'importance'.

We usually prefer a more visual, interactive method, as described in the methods section later in this chapter. Here, we use an Assumption Map to plot each of our assertions on a framework of importance and uncertainty. In this way you can easily see where to focus your efforts by asking of each assumption:

A. How important is it to the project's success?
B. What is your level of confidence that you may rely on this assumption to support the success?

VALIDATING EXPERIMENTS AND PROTOTYPES

Your business proposal is validated when you find data that confirm its supporting assumptions. Remember, we are looking for data with predictive validity, so this adds a complicating layer to our search. Scientists who want absolute confirmation usually resort to 'laboratory conditions' to find an environment that guarantees predictive validity. In the real business world, of course you cannot specify or even know all relevant parameters of the environment. A judicious common sense using reasonable probabilities informed by experience is the only way forward.

DO YOU HAVE ENOUGH CONFIRMING DATA ALREADY?

When you have prioritised your assumptions for validation via the Assumption Map and you focus on each in turn, you should first ask if you already have data that supports the assumption or if such data are easily accessible. It may be that the data are in raw form and require analytical or statistical treatment to extract the information you need. Quantitative analytics can be rigorous and validating. However, you must carefully ensure that conditions in which the data were originally gathered are applicable to the present context for which validation is being sought.

GENERATING YOUR OWN CONFIRMING DATA: VFS

If you do not have validating data available or accessible, you need to generate the data yourself. In particular for Market (Viability) and Technology (Feasibility) questions, there are well-developed and popular methods from general business practice for acquiring confirming data, so we do not describe them in detail here. They include market research, strategic mapping, surveys, competition analysis, financial and margin analysis, business model planning, technology roadmapping, breadboarding, product prototyping and others.

For strategic suitability assumptions' validation, there is only a direct route available. Armed with as much information as you can muster from the other three dimensions (DVF), you must try to convince senior managers to become supporters of the project and draw conclusions from their responses and concerns. Doing this as early as possible is critical. A manager is always likely to be wary of an urgent and unheralded request for support, whereas the manager who has been exposed to and gradually becomes familiar with a project from an early stage will be more comfortable providing assistance or sign-off when required at the right time. Of course, frustratingly, at an early stage a project may be short on supportive, validating data and the manager may need initially to be sold on, or at least introduced to, the dream or an early concept. As usual, the design thinking way is to explore repeatedly with whatever information is to hand in order to learn from responses as early as possible.

Not only senior managers, but also colleagues and partners throughout the firm and externally must be persuaded of the value of the proposal and of the system adjustments necessary to accommodate it.

GENERATING CONFIRMING DATA: DESIRABILITY

Our main focus in this section is to provide guidance on how to go about validating the *Desirability* assumptions of your proposal. As your concept has evolved, many aspects of it are based on your research insights, which came directly from your users. This is a great and essential start, but these features will have been re-interpreted, re-moulded and merged with other features as you evolved the concept. You need continuously to bring your current version of proposal back to the user and get confirmation that its development is proceeding in the right direction. This applies to individual features as well as to the integrated offering. Prototyping is the principal tool that helps you to do this.

Prototyping for concept validation

A prototype is a prop to help you learn. It may be anything that allows the user to get some tangible experience of the new product, service or system being proposed. At earlier concept stages, low-fidelity prototypes and rapid prototypes are the currency with which you build value in your concept because they satisfy the twin goals of high-quality, rapid learning as well as minimal investment of time or cost.

Not only that, but prototypes assist in the creative thinking process. The very acts of designing and constructing the prototype help inform a better appreciation of the concept's strengths and weaknesses and bring to mind new ideas for improvement. In fact, it is an unfair understatement to say that prototypes 'assist' in thinking. As the epitome of visual and sensual thinking, prototypes unfailingly lead the way to new understanding of complex problems and sophisticated yet practical solutions. See Figure 6.5 for a world-famous example in the realms of biochemistry.

FIGURE 6.5   Build to think.

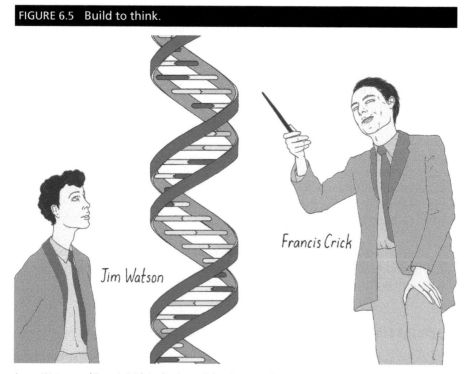

James Watson and Francis Crick in the Cavendish Laboratory (1953) built multiple cardboard and wire prototypes to help them think about the structure of DNA.

Source: © The Explorer's Guide to Biology, Ibiology Inc., with permission.

Throughout history and across many disciplines, people have built to think, and it is a welcome practice that design is now re-popularising this vital skill at the heart of business innovation. It provides essential differentiation from and complement to more established business methods.

When should you prototype?

There are common dilemmas experienced by innovation teams when considering how to prototype their concepts. Mostly, they concern decisions about the comprehensiveness required, the level of detail required and the appropriate timing. The simple advice is that you should prototype regularly, as quickly as possible and as minimally as is required for the purpose at hand.

Let us expand on what this means. Note that throughout any innovation project doubt is constant; it is only substantially relieved after successful deployment, and even then not fully. When the doubt is specific then you should seek out as much learning as is possible at that time to resolve the doubt. It is always a judgement call whether you should divert attention to resolve your doubt through testing a particular issue at that time or whether to carry forward an assumption for more definitive later testing. Of course, this dilemma is more easily resolved if you can devise a quick and inexpensive prototype and test for the issue and moment at hand.

But, doubts are seldom binary yes–no states: will I or won't I do this? Doubts are usually fuzzy, ill-defined, unquantifiable, difficult to comprehend states. Here, you need to sort out your thoughts for your own clarity and sanity, and you need to communicate the outcome as clearly as possible to those who can help you resolve them. Always, the acts of visualisation and making your ideas tangible help you to formulate your thoughts better in your own mind. You should build to think.

LOW-FIDELITY PROTOTYPING

Any visual or tangible artefact that you produce (i.e. a prototype) will crystallise your thoughts and help communicate them much better than words alone. Sometimes it does not take much more than a sketch, a cut-out piece of cardboard, a bricolage of readily available materials and objects, an enactment of a scenario captured in a simple video, or some other product of your ingenuity. These are what we call low-fi prototypes. That is, there is no attempt to simulate faithfully what may be the final end product. After all, you are still in learning mode so you do not really know what the full end product details may be. Tom Kelley of IDEO calls it "mastering the art of squinting", which is to say you are looking for the outline and the essence of the proposal rather than the relatively less consequential, fine implementation details.

In fact, the low-fi prototype actually benefits more by not having too much detail. It enables both the innovator and the user not to be distracted from the core message of the prototype. In addition, prototyping that is deliberately low-fi is often desirable so as to counter a user's naturally sympathetic response behaviour. If you show something and it appears that you have put substantial effort into its production, most people will respond with empathy brimming over to sympathy and they will feel awkward or be reluctant to criticise it, even though internally they may feel it deserves harsher criticism. On the other hand, if it appears to the user that you have not put much effort into its production (notwithstanding the substantial research, reframing and ideation work that you have done but which remain in the background), then that person will more freely give an honest opinion, expecting that you will be less sensitive to criticism because you have not invested so much.

Low fidelity and speed to prototype go hand in hand. But even if there were not the independent merits to low-fi prototypes as we have shown above, speed of prototype execution is extremely important, because the faster you learn about your concept and its context of use, the more productive your continuing innovation efforts will be.

Low fidelity prototyping

According to Tim Brown (2009), "A prototype is anything tangible that lets us explore an idea, evaluate it, and push it forward". A prototype helps represent your concept in ways that words alone cannot. At early concept stages a low-fidelity prototype has some great advantages:

- It is quickly created and becomes useful in minutes.
- The conception and construction of a prototype provides an alternative way to creatively think about and discover new ideas.
- Its lack of refinement means there is little investment in its creation. The innovator is not overly committed to the idea and remains open to learn, and to explore alternatives.

Users who experience the low-fidelity prototype see that is not meant to be a finished article. They will be less inclined to focus on details. Additionally, they will be less likely to be complimentary about it merely to avoid offence to its creator.

PROTOTYPING NEEDS CREATIVITY, INGENUITY, HONESTY

Here is where great skill, ingenuity and creativity may be employed. Can you find a low-cost yet quick-to-implement method that will allow you to get the reliable answers you need within hours or days? The prototype must directly and reliably address your specific questions, concerning the validity of the assumption under study. Sometimes, the target learning will centre around one assumption, perhaps represented by a single concept feature, and the prototype may bear no resemblance to a full concept. It needs only answer the specific question.

At other times, the aggregate of multiple features in a consolidated concept will be represented in the prototype. You must be clear on the particular question(s) you are seeking to answer, and you must be true to your cause and acknowledge and act upon an unexpected response! If an assumption is worth testing, then an unexpected response is worth careful consideration and action, and must not be ignored.

CONTINUE PROTOTYPING, TESTING AND UPDATING

As we mentioned earlier, timing plays an important role in assumption testing. Often, at a certain stage of the project an assumption may not yet be fully testable, for example because all relevant information is unavailable, but you might still value having an initial ballpark check on it. For example, early predictive indicators of assumed profit margin or technical performance may be tested under limited conditions at an early stage, but final performance in margin or technical specification may not yet be determinable. Despite this lack of exactness, sometimes early 'sanity checks' are essential.

Remember, your approach to assumption validation cannot be too linear. In an ideal world you would like to test everything immediately. In practice, the sequencing of your testing must oscillate around the DVFS dimensions of your proposal as well as between definite determination of an assumption's validity on one occasion and an early sanity check indication on another occasion. Then again, sometimes you want to test an individual assumption or feature; at other times you will need to get a fully integrative perspective of all aspects of the proposal as a unit.

What has been the purpose of the repeated prototype and test iterations? Each iteration provides new data to update, refine and de-risk your business proposal. You continue to hone your proposal until you are satisfied that you have diminishing returns of learning from prototyping together with sufficient confidence of risk level that is acceptable in progressing to deployment.

METHODS IN THIS CHAPTER

In driving the innovation from Product Concept to Business Proposal, we capture and prioritise our key assumptions using these methods, which are described in more detail in the following methods section of this chapter.

- Business Proposal: DVFS Assertions Worksheets
- Pre-mortem
- Lean Canvas
- Assumption Mapping

To continuously learn and resolve doubt throughout the concept development and refinement, there is a large armoury of prototyping techniques available, which continue to grow with time. Some are two-dimensional, some are three-dimensional, some are virtual and some are live enactments. A small selection is described below:

- Wizard of Oz simulations
- Role play
- Paper prototyping
- Split testing
- Fake door testing
- Future press release
- Bricolage

Remember, in this chapter and book we continue to emphasise and provide most detail on the *Desirability* dimension of the business proposal, and these methods deal with this alone. Other dimensions (VFS) are treated exhaustively elsewhere in publications dedicated to those topics, and we give references to some of these at the end of the chapter (Bland and Osterwalder, 2020; Blank, 2013; Maurya, 2016; Ries, 2011).

Additionally, always remember that the assumptions you must validate are specific to your concept, your extended business proposal and the context of its use. It may not always be appropriate to replicate exactly some of the techniques we describe here. The possibilities for prototyping and testing to get reliable results are limited mostly by your imagination.

OUTPUTS AND TEMPLATES

At the end of the Validate phase, you have a business proposal that has been validated as far as possible before initial deployment. You have justified that it is adequately de-risked across the four dimensions of desirability, viability, feasibility and strategic suitability.

Your business proposal is represented in summary form by:

- Updated *Concept Board*
- Updated *Business Proposal: DVFS Assertions Worksheets*
- Introductory *Lean Canvas*

Validate | methods

a
audit
What is
known?

r
research
Find out
more

r
reframe
Change
Perspective

i
ideate
New Ideas &
Concepts

V
validate
Test with
users

e
execute
Develop,
deploy & scale

 1

BUILDING THE PROPOSAL
Determine best estimates of a
business proposal across all DVFS
dimensions.

 2

PRE-MORTEM
Has your team forgotten anything
that might cause your project to
fail? Are there hidden weaknesses
in the proposal?

 3

LEAN CANVAS
A 1-page business model template
to deconstruct your innovation into
9 elements, most important for
commercial success.

 4

ASSUMPTION MAPPING
A process for sorting and prioritising
all project success factors under the
DVFS dimensions.

 5

WIZARD OF OZ SIMULATION
A service or experience is delivered
to a user, which provides illusion of
full implementation.

 6

ROLE PLAY
Acting out a future scenario of a
customer journey, for the purpose
of empathy and understanding
experiences.

 7

PAPER PROTOTYPING
An excellent, quick, cheap way to
test and evaluate key desirability
assumptions of early stage
concepts.

 8

OTHER TESTING METHODS
- Split Testing
- Fake Door Testing
- Future Press Release
- Bricolage

Validation is about de-risking the project. Identify key assumptions in a concept and business proposal, and then test them and amend them with repeated experimentation.

Go beyond desirability to build a full proposal for your concept.

You have developed concepts as integrated value proposition offerings for the user, with desirability as the guiding mantra. In order to make a business case for internal-to-the-firm evaluation and investment decision you must build a business proposal from the concept, which adds detail around market, technology, financial and strategic suitability. There is a wide selection of established techniques for researching and designing business models, and conducting market research, financial analyses and technology evaluations. Most firms will have their own methodologies, and detailed consideration of these is beyond the scope of this book.

We deal here with the initial fleshing-out of the business proposal, which provides a basis for incrementally improving its accuracy through deeper research and validation experiments.

Though user desirability has been the main emphasis up to now, you and your team have not ignored the VFS dimensions. (Refer to the FVS Sanity Index of chapter 5). Turn your attention now to apply this knowledge, experience and intuition in order to build an initial estimate of business proposal.

We use four pro-forma worksheets to assist this process, recognising that many of the questions to be considered are generic. Where there are specific considerations for your project, of course these may be added easily, and you should build a proposal that suits your specific project and firm requirements.

Inevitably, there will be refinements or even significant amendments to the product concept, which will be mandated by the extended VFS considerations in the proposal.

Step by step

1. **Convene the meeting**

   Assemble the team for a three-hour meeting. Prepare the room with large displays of the Reframed Project Vision, Product Concept Boards and the *DVFS Assertions Worksheets* shown overleaf.

   Remind the team that the ultimate objective of the project is to deploy a product (service/system/offering) that will sustainably succeed in the market. The core concept has been the main focus up to now and is the heart of the offering. It needs a well-refined structural and support framework (VFS) to achieve scale.

2. **Work in detail through the worksheets overleaf**

   Refer to the *Desirability* worksheet first, in large display on the wall or projector. Discuss each prompt in turn and note the agreed response (assertion). Responses must be *succinct* (use short, clear sentences); *to-the-point* (without obfuscation, exaggeration or ambiguity); *assertive* (no wavering or conditionality – take a clear stand).

   Desirability should be the most straightforward worksheet, given it has received primary attention up to now, but it can be surprising how team members often retain different understandings. Try to reach agreement. If there are points of sustained discussion or disagreement, note these and move on. Keep moving. Repeat for the other worksheets. Add or amend the worksheet topics as you feel appropriate for your project and firm. Where there is a high level of uncertainty about a topic, put it aside for investigation over the next few days, possibly with the help of relevant expert opinion.

3. **Complete the worksheets**

   You may need a second meeting to tidy up loose ends. Do this within a short timeframe. Remember, the worksheets are your best current estimate of a coherent proposal. They are not 100 per cent definitive; they contain many assumptions that need to be formulated as hypotheses and tested. In effect they are the working brief for Validation. The Assumption Mapping, later in this chapter, prioritises the validation efforts.

BUILDING THE BUSINESS PROPOSAL templates 2/4

FIGURE 6.6 DVFS Assertions – Desirability worksheet.

	Assertions worksheet to build a Business Proposal	1. DESIRABILITY (add or delete to suit your own project)	
D1	Target Customers will be:		
D2	Customers will use it to do this/these job(s):		
D3	Obstacles to successful completing the job(s) are:		
D4	Customers will measure outcome success by:		
D5	Customers will prefer it over these alternative solutions, for these reasons:	*Alternative solution*	*Reason why our solution is preferred*
D6	Customers will be compelled to switch to our solution, because of this overwhelming advantage:		

FIGURE 6.7 DVFS Assertions – Feasibility worksheet.

	Assertions worksheet to build a Business Proposal	3. FEASIBILITY (add or delete to suit your own project)	
F1	Our significant technical or engineering challenges are these, and we will overcome them in-time/on-budget in this way.	*Technical Challenge*	*How we will overcome the challenge*
F2	Our solution is technically superior to alternatives in this way:		
F3	We're confident our team is technically competent for this development job because:		
F4	Our development time to deployment readiness is:		
F5	Our funding amount required for technical development is X and we will secure this funding from Y.	*X*	*Y*

BUILDING THE BUSINESS PROPOSAL example 3/4

FIGURE 6.8 DVFS Assertions – Viability worksheet.

Assertions worksheet to build a Business Proposal	2. VIABILITY (add or delete to suit your own project)		
V1	We will achieve growth to X in time Y, because:			
V2	We will acquire new customers by these means:			
V3	Our primary competition will be:			
V4	The competition will respond to our solution in this way:			
V5	We will generate revenue in these ways:			
		Year 1	Year 2	Year 3
V6	New business from referrals will be [%]			
V7	Repeat business rate will be [%]			
V8	Our quantity of paying customers will be:			
V9	Our average revenue per customer will be (£/$/€):			
V10	Our average gross margin will be (%):			
V11	We're confident our Operations organisation has competence & capacity to deploy this project because:			
V12	The project 3-year investment numbers will be:	Amount	NPV	IRR

FIGURE 6.9 DVFS Assertions – Suitability worksheet.

Assertions worksheet to build a Business Proposal	4. SUITABILITY (add or delete to suit your own project)
S1	Our firm leadership will support this project, because:	
S2	This project enhances the firm's brand, because:	
S3	The scale and scope of this project fits our organisation and its mission, because:	
S4	This project will be prioritised sufficiently to attract resources for development, because:	

Q What is the difference between the terms assertions, assumptions and hypotheses?

A We are chasing a subtle but important emphasis when we use the term assertion, which goes beyond different meanings in a dictionary. Consider this:

You are walking in a field and you see apples on the ground under an apple tree. You might observe:

(A) "The apples fell from the tree."

or,

(B) "I assume the apples fell from the tree." ... by which I mean this is more likely than they were left there by the apple pickers who plucked them from the tree, but this alternative is still a possibility!

The former is an assertion of fact. The latter is an assumption that explicitly allows for alternative possibilities.

An assumption formulation is more correct for most things. However, an assertion has a tone (of confidence, even arrogance) that invites disagreement, and may be directly challenged and tested. It is this 'call to action' tone that we like. It crystallises the thought and helps to define a hypothesis and test that will determine its validity. Later in the chapter, we revert to the more accurate term 'assumption'.

We reserve the term 'hypothesis' for the specific formulation of an assumption, or part of an assumption, that will be directly tested.

Q Our team cannot know all of the business and market data that are called for in the worksheets. What should we do?

A They cannot be experts in everything but, probably and desirably, team members will have a good, broad understanding and expectations. A project team cannot operate in isolation within the firm, because "no team is an island", to paraphrase the poet John Donne. Additionally, after immersion in the topic for weeks or months, there will inevitably be a good appreciation of the broader business parameters, even unconsciously. This must be captured.

Remember, the assertions are not meant to be definitive and 100 per cent perfect. They are meant to be the best estimates of the moment. They form a base that will be further rigorously tested in Validation phase and early launch.

You could consider asking a representative from Marketing, Finance or R&D Engineering to join your team for the meetings, but be sure to brief them on the purpose and desired outcome of the sessions. As well, if a topic is undecided after the first meeting, consult with such domain experts for input and guidance.

PRE-MORTEM method   1/4

Change your perspective from hoping for success to expecting failure, and discover hidden weaknesses in your proposal.

We believe in positivity and the power of positive thought. But, we also recognise self-delusion and biases of convenience and familiarity, and others.

The business proposal you drafted in the previous method contains assumptions with varying levels of plausibility and likely validity. Soon, you will prioritise these for testing using the Assumption Mapping method we describe later. First, we encourage you to consider the whole proposal using a very different mindset. How might it go wrong? When you explore with this (negative, critical) mindset you can uncover new issues that are not highlighted in a standard framework like the DVFS worksheets introduced earlier.

A pre-mortem uses prospective hindsight, which increases by 30 per cent an individual's ability to correctly identify reasons for future outcomes (Klein, 2007). The pre-mortem asks an innovation team and other knowledgeable stakeholders to imagine themselves in the future and to posit that a project has failed, either marginally or drastically. The group then brainstorms to come up with reasons as to why it might have failed. It is important participants are encouraged to include any reason whatsoever, no matter how eccentric, politically incorrect or contrary to perceived wisdom.

In a second round, the actions or means to mitigate each failure mode are considered and a solution is asserted for each. These become additional conditions for success of the project, which may be added to those already identified under the 'assertions' method of the last section. In a workshop, these new assertions are captured in Post-it notes.

Step by step

1. Convene the meeting

Assemble the team for a one-hour meeting. Prepare the room with large displays of the Reframed Project Vision, Product Concept Boards and the *DVFS Assertions Worksheets* from the previous section.

Remind the team that the project is successful only when the product (service/system/offering) is deployed successfully and sustainably in the market. You hope you are on the right track! But, what if you are not? Have you forgotten anything? Is there anything the team has not considered that could cause the project to fail?

This is the place to think divergently, even eccentrically. Is the firm not ready for it in some way, due to exhaustion after hard times or arrogance after good times, for example? Is there boardroom tension that will get in the way of a reasonable consideration for the proposal? Hopefully, your team meetings are 'safe places' to bring these concerns into the open. Note that only the later proposed mitigation actions need to be brought outside the room.

2. **The project failed because ...**

Working individually, ask team members to write all possible reasons for failure on their individual notepads. This should take five minutes. Then, working in sub-teams of 4–6 people, transfer all individual reasons for failure to form a list on a flipchart or whiteboard, for all to see. Discuss freely and add further reasons as they occur. [20 minutes]

3. **Mitigating the failure modes**

Now, go through the list one by one and ask:

"What actions might we take, which we assume will mitigate the reason for failure?" [20 minutes]

Collect each action or proposal on a separate Post-it note. These Post-its become additional conditions for success of the project, which are added to those already identified under 'DVFS Assertions'.

PRE-MORTEM template

FIGURE 6.10 Pre-mortem process.

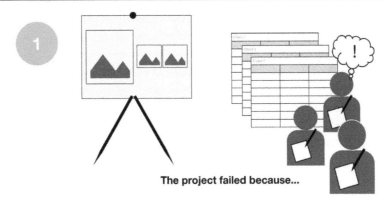

The project failed because...

FIGURE 6.11   Pre-mortem example output.

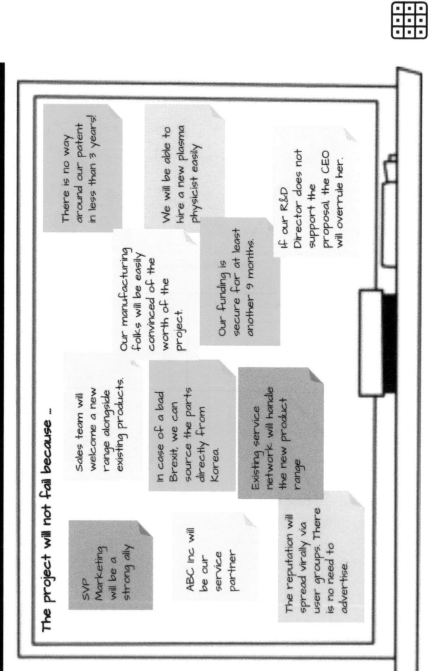

The project will not fail because ...

SVP Marketing will be a strong ally

ABC inc will be our service partner

The reputation will spread virally via user groups. There is no need to advertise.

Sales team will welcome a new range alongside existing products.

In case of a bad Brexit, we can source the parts directly from Korea.

Existing service network will handle the new product range

Our manufacturing folks will be easily convinced of the worth of the project.

Our funding is secure for at least another 9 months.

There is no way around our patent in less than 3 years!

We will be able to hire a new plasma physicist easily

If our R&D Director does not support the proposal, the CEO will overrule her.

PRE-MORTEM faqs

Q Have we not given this enough thought? Why waste more time?

A Would it not be great if this were the only hour that you wasted in your project? The sad reality is that no concept or proposal will ever be perfect and you can never accurately predict users' and other stakeholders' future behaviours.
We believe it is better to spend an hour now probing for possible weaknesses that may be mitigated in advance, rather than scrambling for many hours, days or months at a later stage trying to fix things on the fly, if it is not too late.

Q It feels quite negative; we are acting like doomsayers. How can we prevent it from dispiriting the team?

A We have never found this to be a problem. Remind the team they are transporting themselves into a 'special frame' for just an hour.
In addition, by making it into a fun challenge you will encourage participants to use their ingenuity actively to seek out areas of greatest weakness, instead of applying the same ingenuity counterproductively in justifying to themselves and others why the possible weaknesses may be covered already.
Everyone loves a challenge!

Q Should we include participants who are not regular team members?

A Most certainly, yes. Every firm has people who seem to specialise in scepticism, even cynicism. So, by all means, bring them in and let them apply their special talents!
The only condition, as always, is to ensure they are properly briefed. Especially, they must acknowledge and accept the new-born and un-validated nature of the proposal. Additionally, they must respect the exploratory nature of the exercise and respect the safe space and mutual trust that is necessary for that.

"A problem well stated is a problem half-solved."

Charles Kettering

The Lean Canvas is a tool that supports the design and iteration of a new business model using Lean Startup principles.

A business model describes how an organisation creates, delivers and captures value. The Lean Canvas consists of a grid of cells, representing nine essential components of a business model. It is especially suited to guiding startups and innovation initiatives facing market uncertainty.

Originally created in 2010 by Ash Maurya (2016), the Lean Canvas was a development of the business model canvas created earlier by Alexander Osterwalder. For Maurya, the Lean Canvas variation was better attuned to the special conditions of startups and entrepreneurs, compared to established businesses, and we agree.

The Lean Canvas serves as a prototyping tool by allowing an entrepreneurial or innovation team to configure and visualise a full business model, and then review it and submit it to testing. Its key advantage is the business model may be grasped in overview, with an integrating narrative, and considered holistically. Initially, the 'prototype' testing may be through systematic thinking and team discussion, with the canvas as a prop. Later, field experiments are used to evaluate and validate isolated assumptions within the business model and eventually to track progress of the business model's evolution.

The canvas is usually iterated many times. To facilitate a compelling narrative of the story behind the business model, we describe here our preferred sequence for populating and narrating the canvas.

Step by step

1. **Convene the team meeting**
   The purpose of the meeting is to draft a complete business model that supports your new concept. Prepare the room with large displays of the Concept Board and DVFS worksheets. A particularly large version of the Lean Canvas must be used. This may be A1 or A0 size poster, or it may easily be drawn on a whiteboard.

   Build a narrative of how the business model works, starting with the customer segments, and concluding with the unfair advantage or distinctive differentiation that will cause your project to be a commercial success.
2. **Populate the canvas**
   The facilitator populates the canvas with Post-its in response to discussion and agreement by the team. Work around the canvas as shown on the next page, paying attention to the prompts in each section. In the event of disagreement, discuss for some time but do not get bogged down. Keep moving, and note where large disagreement or doubt exists.

3. **Check for functionality, coherence and ambition**

   Review the populated canvas. How does it feel? Will it work? Is it sensible and coherent? Is it ambitious enough? How can it be improved? Revert to the noted points of disagreement and discuss them in greater detail now in light of the emerging bigger picture.

   In many cases it makes sense to explore two or more separate business models and probe their relative merits through validation testing. Be creative and ambitious. Do not settle too easily for a standard approach.

4. **Identify assumptions**

   Identify the key assumptions associated with each business model. How might these be tested?

LEAN CANVAS template

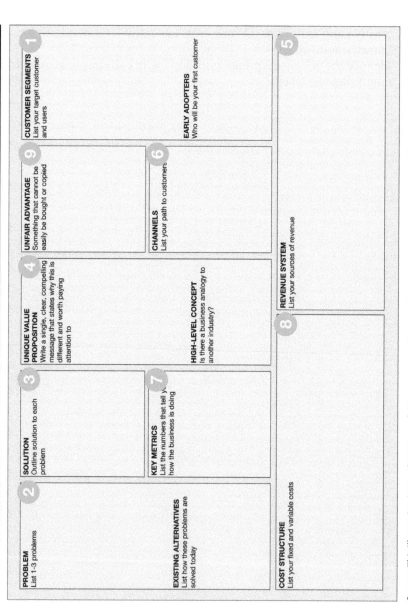

FIGURE 6.12 Lean Canvas template.

CUSTOMER SEGMENTS 1
List your target customer
and users

EARLY ADOPTERS 5
Who will be your first customer

UNFAIR ADVANTAGE 9
Something that cannot be
easily be bought or copied

CHANNELS 6
List your path to customers

**UNIQUE VALUE
PROPOSITION** 4
Write a single, clear, compelling
message that states why this is
different and worth paying
attention to

HIGH-LEVEL CONCEPT
Is there a business analogy to
another industry?

REVENUE SYSTEM 8
List your sources of revenue

SOLUTION 3
Outline solution to each
problem

KEY METRICS 7
List the numbers that tell y...
how the business is doing

PROBLEM 2
List 1-3 problems

EXISTING ALTERNATIVES
List how these problems are
solved today

COST STRUCTURE
List your fixed and variable costs

LEAN CANVAS example 3/4

FIGURE 6.13 Lean Canvas example.

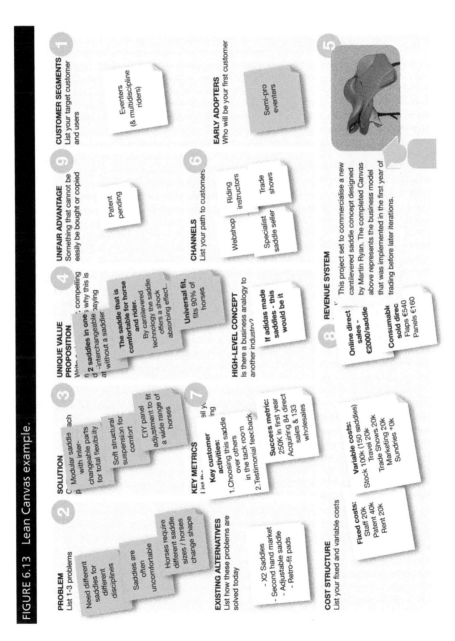

Source: Cantilevered Saddle by Martin Ryan.

Q Can I complete a Lean Canvas before the concept is finalised?

A Yes. While we have introduced the Lean Canvas now, as a method of Validation, it provides an excellent template for thinking more roundedly about a value proposition at any time. It starts with the problem and target customer and encourages the innovator to think more clearly about the underlying rationale supporting the project and solution. At any time of the project, this is a helpful exercise.

If completing a Lean Canvas while the concept is still incomplete, you might consider to omit sections 5, 6, 7 and 8.

Q Is there more information available about the Lean Canvas?

A There is a large number of books, and professional and academic papers written about the Lean Startup methodology and the two most popular canvases (Business Model Canvas and Lean Canvas). We have referred to some of these in the bibliography. For Lean Canvas, we especially recommend the book *Scaling Lean* by Ash Maurya (2016).

Q How much detail should go into each section?

A Keep it high-level. You do not want to get bogged down with detail, most of which is unknowable at this stage anyway. You are looking for the overall configuration. As Tom Kelley of IDEO says: "Master the art of squinting".

Focus on one customer segment for each canvas. Choose the customer segment that is most likely to lead to a successful business outcome. If you want to tackle more customer segments, then complete more canvases. Information should be sufficiently succinct that you can quickly appraise the essentials of your business, without getting lost in heavily detailed reading.

It can be hard to write points simply and succinctly, so this may take some work and time to tease out, but it gets easier with practice.

ASSUMPTION MAPPING method   1/4

Why will the project succeed with this business proposal? What are the factors that are critical for it to succeed, and how can we prioritise them for investigation?

You have identified many factors that are necessary for success, using the DVFS Assertions and pre-mortem methods. You expressed the factors as assertions, in order to create a tone that invites interrogation and testing. While building the full proposal you assumed the assertions are true, in order to proceed with a comprehensive configuration.

But, in reality, the assumptions have varying degrees of uncertainty, ranging from fully proven to unfounded speculation. Further, it is clear that the various assumptions also have different levels of importance, which relates to their contribution to success of the project. Some factors are unique and irreplaceable, while others have easy alternatives with no negative impact on the project's budget or schedule.

These two parameters – uncertainty and importance – provide a useful basis for sorting the assumptions and for prioritising them according to urgency for testing and contingency planning.

Assumption Mapping prepares for fast testing and project de-risking, by identifying the risk factors most in need of mitigation at any time.

> "Win fast,
> win slow,
> lose fast,
> but never ever lose slow."
> > John McCormack, Microsoft Ireland

Step by step

1. **Identify and list all risky assumptions**
   Gather the team for an approximate one-hour meeting. Display the DVFS Assertions and the Pre-mortem Actions. Draw the axes for a large Assumption Map on a whiteboard, using the template on the next page.
2. **Place one assumption at a time**
   Take one assumption at a time from the DVFS Assertions and Pre-mortem Actions collection. Call it out loud for the team as you transfer it to a Post-it, which you place on the whiteboard map in a location agreed with the team. Agreement may not come easily! Allow for some, but not too much, discussion and then the facilitator makes the call where to place the Post-it. Remember, we use Post-its so that they may be moved later.
3. **Repeat for all assumptions**
   Note well: Maintain a good spread amongst the Post-it placement locations. Considered individually, everything must be relevant or it should not be there! In practice, however, all things cannot be treated as equally urgent. Constantly evaluate their relative positioning and move previous placements as necessary.

4. **Plan testing and contingency schedule**

Assumptions that fall in the top right quadrant are your priorities for testing. These are most important for project success, yet are also the most uncertain. Plan tests that will reduce uncertainty and risk, as quickly as possible.

Others that are close to or overlapping this quadrant must be evaluated in more detail and, perhaps, contingency courses of action may be planned.

ASSUMPTION MAPPING template

FIGURE 6.14 Assumption Map template.

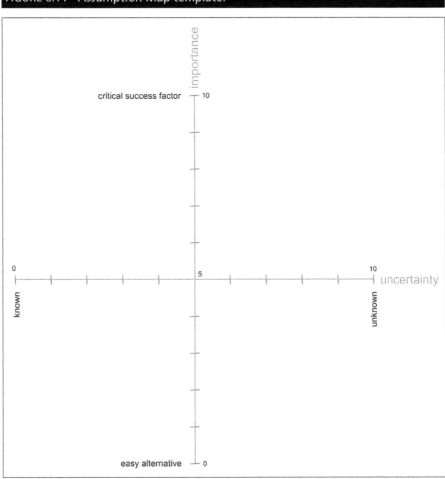

FIGURE 6.15   Assumption Map example.

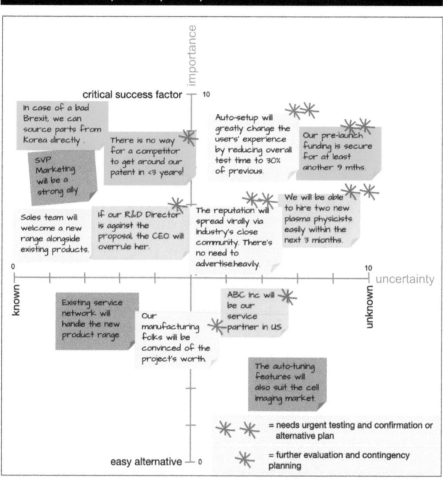

Q The Post-it is physically big relative to the map. How can we place it accurately?

A Fine accuracy is impossible at this early concept and/or business proposal stage. It would be a folly to delude yourself that you could be accurate in regard to such a difficult prediction of a complex future. So, the relatively large size of a large Post-it represents this inaccuracy nicely.

Further, the main objective is not to determine an absolute risk level for any assumption (impossible). The Assumption Mapping is more for the purpose of prioritising, i.e. which of the assumptions embodies more risk for the project than others? These are obviously the ones that must be given most urgent attention.

Q It is hard to figure out the scaling of each axis. How can we agree this?

A The scale is defined by the end points and mid-point, where the axes intersect. We use the following rubric:

Importance
 0 = alternatives easily available
 5 = workarounds available; impacts on project time and budget.
10 = critical success factor; no alternative

Uncertainty
 0 = known; fully evidenced
 5 = moderate confidence; some supporting evidence
10 = unknown; no direct evidence

Q How should we test our urgent assumptions?

A As quickly, effectively and cheaply as possible. In the next few method sections, we share a number of testing approaches, particularly suited to the desirability dimension of the proposal. Our methods are versatile approaches that can be effective across different types of value proposition.

For VFS dimensions, there are common methods from general business practice that you might use, which are well developed and often applied, but they are outside our scope and we do not cover them here. See Bland and Osterwalder (2020).

We emphasise qualitative methodologies but of course quantitative and mixed methods are useful, too. However, be aware that quantitative methodologies are often unsuited to early stage testing as the necessary variables are too ambiguous or under-defined.

WIZARD OF OZ   method   1/4

> "The only way to engineer the future tomorrow is to have lived it yesterday."
>
> Buxton, 2007

A Wizard of Oz experiment is the practice of illusion. An effect, service or experience is delivered to a user with a manual, ad-hoc delivery system that provides illusion of deeper, systematic implementation.

The Wizard of Oz was the title of an iconic 1939 film based on the popular 1900 children's novel by L. Frank Baum, *The Wonderful Wizard of Oz*. In the story, the powerful and mysterious Wizard of Oz dispenses wisdom and advice. Towards the end of the story, it is revealed that the Wizard is nothing more than a humble old man behind a curtain. However, to his visitors he had appeared as a wise wizard.

In modern times, the concept was first used in software testing to indicate giving an impression that a fully automated software application exists, when in fact a human operator is manually providing the intelligence and functionality. The software interface acts like the curtain in the Oz story, and the software user does not see behind it. The user experiences the service as if it were real, while the software designers get reliable and valuable feedback without the time and expense of creating a working system.

Today this methodology is used more broadly to test physical, digital, and service offerings. The innovation team must employ creativity and ingenuity to 'fake' the solution into existence, at minimum expense.

In planning your Wizard of Oz experiment we recommend you satisfy three criteria:

1. The user experience must appear sufficiently real to stimulate a realistic reaction from the test subject.
2. Test early, as soon as you have some understanding of the desired experience from your concept.
3. The test should be cheap and quick, having only minimal viable fidelity to serve its intended purpose. Maximum learning from minimum expenditure and effort is the motto.

Step by step

1. **Identify and clarify the assumption to be tested**
   Identify the assumption you wish to test. Analyse it and break it down to its core, removing any features that may be peripheral or unnecessary. What is the intended stimulus and expected outcome? What is the context? Who are the people involved? What are the essential interactions? What is the desired experience of the user?

   Note, as we have mentioned previously, that here we consider only those assumptions that relate to the 'desirability' dimension of your concept. Wizard of Oz is especially suited to these.

2. **Plan how to simulate the desired user experience**

    Using any ideating technique(s) you may choose, devise an experiment that will deliver the intended experience to the user with the minimum expenditure of time, money and effort. Pay attention to all factors that are relevant to delivering the experience. Ideally, the experiment will appear as a fully authentic experience for the user. Alternatively, where that is not possible, the user may be aware of the experimental nature of the experience but may be sufficiently absorbed in the experience to appreciate what it could be like in reality.

3. **Create test props and assign roles**

    Create necessary props, images, videos and animations to support the experimental set-up. Assign experimenter roles as 'wizards', actors and observers.

4. **Conduct the experiment**

    Conduct the experiment as often as necessary or convenient. Have the designated observer (and all those involved, where possible) take notes, photos, videos, etc. Note what works and what does not, and where the moments of frustration are. After the experiments, ask the *users* about their impressions in the form of feedback and advice for improvement.

FIGURE 6.16 An illustration of the passage from *The Wizard of Oz* (MGM, 1939), where the wizard was finally revealed from behind the curtain. Up to that point he created the illusion of a great and powerful wizard.

WIZARD OF OZ   example   3/4

FIGURE 6.17   Wizard of Oz example.

One of the most downloaded children's apps, Elmo's Monster Moves designed by IDEO Play Lab, used a 'Wizard of Oz' illusion to demonstrate a working mobile interface that didn't exist. A simulation of the experience was created using some card props and members of the team to act out a use case, while a recording was made of just the screen area.

Q How can I be sure that my experiment will work, before going to the expense of setting it up?

A You cannot be sure. At early concept stages in particular, parameters are fluid and it can be difficult to know what questions to ask. Notwithstanding this, from building your concept and business proposal as described at the beginning of this chapter, your initial assumptions should have some definition. With good critical thinking, you should be able to hone your understanding. But, in the midst of fluid uncertainty, you cannot think your way to a conclusion. You must act.

Hence, we say, your experiments must be fast and cheap (low cost). You always want to get maximum (learning) impact for your efforts. If you work creatively, you do not need to spend much time or effort and your question becomes irrelevant.

Q I accept 'as little effort as possible'. But how much is that, typically?

A It depends … There are many situations where you could identify an assumption or two in a morning, plan and prepare the experiments through the afternoon, and conduct the experiment the following day.

Q Should I conduct a separate experiment for each assumption or may I combine many assumption tests in one experiment?

A When your experimental arrangement is set up with users, it makes sense that you try to carry out multiple experiments, either simultaneously or sequentially. You might design the experiment deliberately to explore multiple assumptions. But, be careful. More parameters make it more difficult to interpret the outcome. Imagine you had a user who is delighted by an offering. If this offering has been pared down to one significant distinctive feature, then it is reasonable to attribute the favourable outcome to this distinctive feature.

On the other hand, for example, if the offering has three distinctive new features, how would you know to which feature you should attribute the successful outcome. Should you credit feature A or B or C, or A+B or A+C or B+C? Would A+B give a better result if C were not there? It is difficult to interpret such a result unless each assumption has been evaluated previously. In quantitative research, these are called confounding factors.

On the other hand, again, it is often not possible to isolate one feature absolutely and still to retain a realistic simulation. Your best judgement and ingenuity are required.

ROLE PLAY method   1/4

> "The best way of successfully acting a part is to be it."
> 'The Adventure of the Dying Detective', Arthur Conan Doyle, 1913

Act out or perform today a scenario you are planning for the future, so as to empathise with characters and understand their probable experiences.

Role play seeks to place the team in the shoes of the end user or in the mind of the user experiencing a scenario. It affords a first-hand understanding of the prospective experience of participants in a proposed solution, and helps to unlock early signs of weakness and to identify particular strengths. It can offer an immediate and readily accessible validation exercise, delivering deeper understanding limited only by the team's imagination and willingness to act out a situation. The objective is to make an idea tangible enough to provoke a meaningful response.

Role play is particularly relevant to any innovation encompassing a process, a customer journey or activity involving a number of interactions between people and/or artefacts.

Step by step

1. **Select an idea and set out your test objectives**
   Determine which of your ideas you want to role play and why. Decide the test objectives. What are the key assumptions and project risks that you are trying to understand, test and mitigate? Prepare a short list of questions that you plan to answer.

2. **Plan the role play**
   What are the most important inflection points in the journey? Take about 30 minutes to determine the most important steps of the service, process or experience. Decide on the different characters and assign roles. Nominate a team member to play each role.

3. **Gather props to make the experience more real and tangible**
   Embrace 'bricolage' – the practice of constructing something from a diverse range of convenient items. Bring together some essential props or substitute objects that, with a little imagination, can help to make the role play more tangible.

ROLE PLAY template 2/4

FIGURE 6.18 Role play scene preparation.

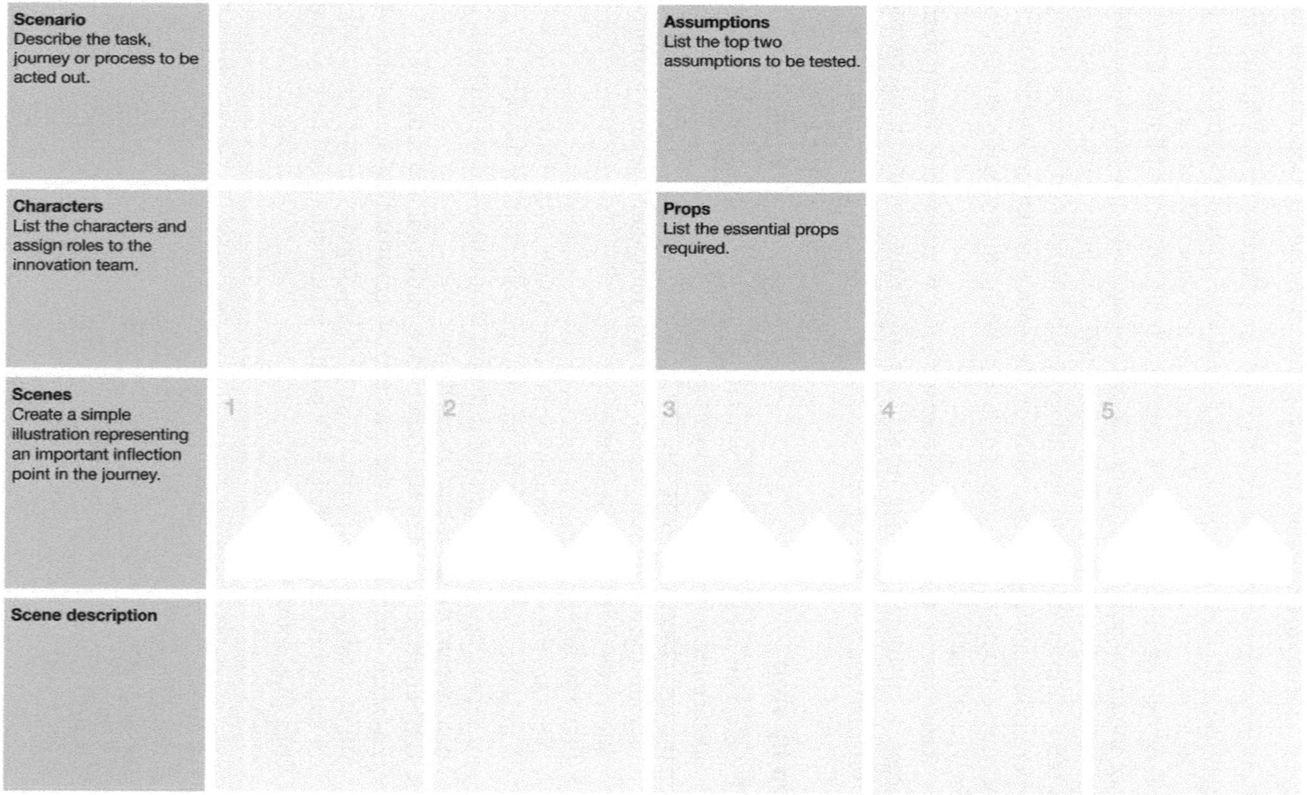

Scenario
Describe the task, journey or process to be acted out.

Characters
List the characters and assign roles to the innovation team.

Assumptions
List the top two assumptions to be tested.

Props
List the essential props required.

Scenes
Create a simple illustration representing an important inflection point in the journey.

1 2 3 4 5

Scene description

ROLE PLAY example

FIGURE 6.19 Role play example.

Two product designers at Maynooth University (Sam Clarke and Aaron O'Neill) are role playing the operation of a new agri-tech solution that works to extract dangerous gases from the surface of slurry tanks on farms. Here the concept was being explained to a designer who worked at Dyson. Interestingly he was totally unfamiliar with this issue on farms, having grown up in a city. Through role play the students were able to impress on the Dyson designer the potential benefits of their solution.

ROLE PLAY faqs

Q Will this work with team members who are shy or cannot act?

A There are two broad types of test scenarios that might benefit from role play. One is an acted-out scenario and journey in front of prospective customers or key stakeholders so as to provoke an experience and garner their feedback. This may even be recorded for sharing with wider networks. The second involves the smaller circle of the innovation team alone. The purpose here is to help the team think through and refine a proposal ahead of external testing. You may need more extrovert and confident actors for the former. However, normally, all team members will be at ease with the latter and they should be reminded that it is an important learning tool for the actor and not just the audience.

Q Are props really necessary?

A Most of the time, yes. However they do not need to be extensive or sophisticated. Use judgement. When testing a service, do not waste time creating a mock-up of an environment if you can easily go to the real thing. Equally, do not make it more complicated than it needs to be. Try the role play with fewer props first and only make further efforts as required. Be careful not to let 'prop perfection' become an obsession and hence a barrier to the real objective. 'Satisfactory and done' is much better than unattainable perfection.

PAPER PROTOTYPING   method   1/4

This is an obvious, yet excellent, way to quickly test and evaluate key desirability assumptions of early stage concepts.

Paper prototyping has a great capacity to make the abstract tangible within hours or even minutes. When you are seeking approval or advice of your prospective users for your concept, you must ensure they understand the proposal fully. Given that most people struggle with comprehending a concept on the basis of verbal description alone, a three-dimensional physical prop or sketch model closes the gap of understanding and militates against misinterpretation.

A paper prototype strikes a balance of three-dimensional clarification and low-fidelity speed. The concept's essential features are made clearer, yet clearly the artefact is merely suggestive, a tentative proposal, which encourages viewers to be both accepting and openly critical where necessary. In contrast, a high-resolution, more polished and complete prototype may inhibit honest feedback as viewers are reluctant to criticise a 'finished' product that obviously took effort to produce.

Paper prototypes are used by every type of firm and project team, from startups to leading corporations. For example,

Digital interfaces: Many of the world's top software apps are first embodied as paper prototypes.

Physical products: The billion-dollar vacuum cleaning manufacturer, Dyson, always uses paper prototypes as an important staging point when iterating its next generation of appliances.

Services and processes: The largest medical device firm, Medtronic, uses card modelling in testing new surgery layouts and configurations with surgeons. In addition, card layouts at full scale are used for testing new manufacturing cell layouts with production operators.

Step by step

1. **Paper prototyping is ideal for rapid co-creation**
   Always be with or near the user. Paper prototyping is especially useful when you are co-creating with a user. A ten-minute mock-up might elicit a useful response from an attending user. If encouraging, this can be built upon immediately with further iterations.

   Where a user is not in the room with the team, be sure to have quick access, so as to get rapid feedback on your efforts. This might be achieved by having the session on site with the user, in the context of use, or with the user virtually present by video-conference, or other means.

2. **Select a concept or feature to be tested**
   Focus on the concept to be validated and its essential features and core value proposition. Identify the key assumptions you wish to test.

3. **How is this assumption to be embodied?**
   How do you envisage the solution working, giving rise to the assumption in focus? What will be the main recognisable features, from the user's perspective?

As the team thinks about and discusses these points, use white-board, wall or flipchart sketches to illustrate and iterate the various points. Encourage team members to use impromptu card or paper representations and so add a three-dimensional aspect to discussions. Iterate using more precisely cut shapes and different coloured or annotated card. Add in other props that may increase the validity of the artefact. However, only use what is conveniently available, so that a long delay does not interrupt the creative flow.

4. **Get rapid feedback**

By team agreement, conclude the evolution of the cardboard proto-type when it has reached a useful representation of the concept and the assumption(s) that are your immediate focus. Let the user(s) touch, hold and manipulate the artefact so as to give them a deeper understanding of what it is and its possibilities.

PAPER PROTOTYPING template

FIGURE 6.20 Paper prototyping essentials.

FIGURE 6.21 Paper prototyping examples.

Q When might I choose paper or card over other materials?

A Paper and card are widely available and easy to work with. They are excellent for communicating form, scale and feature arrangements, and we would generally advise that you try paper or card first for these purposes.
Paper has the benefit of being easily manipulated and readily available. When you show it to a test user and discuss it, it conveniently provides a surface to write and sketch on. Suggested improvements and feedback notes can be directly documented on to the prototype itself.

Q Which do you prefer, paper or card?

A We use the words almost interchangeably, as they can both be bent, glued, coloured and cut with a scalpel. Use card over paper where some extra structural stability is required.

Q How much time would you allocate to paper prototyping?

A For concept evolution, we do not like to think of it in this 'transactional' way. Paper prototyping and sketching should be an everyday tool and activity to help you think, co-create and get immediate feedback on progress. For most product and service concept developments paper prototypes would be created within ten minutes up to an hour or two, and iterated repeatedly with rapid feedback.
Exceptionally, some prototypes may be large-scale, more complex and built to scale for specific measurements of work-flow and ergonomic factors. These are accurately and precision crafted to proportionally represent a product or building arrangement, for example. They require a higher level of refinement and skill and will take longer to construct.

There are dozens of useful, cheap and fast approaches to help you determine if you have a 'problem–solution' fit. Here are some more to finish our selection.

1. Split testing

Also known as A/B testing, split testing describes an experiment where two alternative variations of a solution are created and presented to equal cohorts of the same group. The hope is that the users will show a clear preference for one variation over the other and thereby help determine their preferred configuration of the concept.

The experimenter should have a well-defined user goal, and a clear success metric with which to evaluate the performance of each variation presented. The test works particularly well in website, process, or app design where a small number of elements of the system are under study but the majority of the system elements stay constant. Metrics may be associated with user behaviour, satisfaction, confusion, and willingness or ability to progress through a process, or many other possibilities. For web applications in particular, large amounts of quantitative data may be easily captured. Follow-up interviews help to clarify the underlying cause of the observed behaviours.

To complete a split test:

1. Decide the specific user goal.
2. Create the two variations.
3. Decide on the success metric and means of measurement.
4. Capture results.
5. Follow up with user interviews.

2. Fake door testing

This is a generic term for a method of measuring market interest in your proposed solution before incurring full project investment costs to create it in full. It is similar to the Wizard of Oz method in that it is illusory to the user. However, in the Oz method, the user actually receives a service and, presumably, may be satisfied. Here, the user is seduced into believing there is a service available but, if interested, will find that it is not available and is thus disappointed. Hence, although it is a quick and simple way of getting user response, you must be careful not to disappoint and perhaps disillusion the very users who are interested in your service. They may feel betrayed and suspicious of further engagement with you. You must plan, proactively, how you will handle this disappointment of the prospective user.

Examples:

1. A web landing page that includes outline description and visuals of your offering, with click-through buttons for further information. A short, sharp promotional campaign directs traffic to the page.
2. A professional quality hard-copy brochure, as above.
3. A 'Kickstarter' funding page, with full visuals and quasi-realistic video.

3. Future press release

All innovation efforts must be sold to colleagues within the firm as well as to external customers. In an effort to clarify a business proposal's vision and make it more tangible to colleagues, Amazon's Jeff Bezos requires new proposals to be written into a future press release. The exercise is particularly good at clarifying the killer feature and organising layers of the value proposition. It forces the innovators thoroughly to examine key features, reasons for user adoption, and the project's likely path to success.

To complete a future press release:

1. Set the press release at a future date. Write it to reflect the success expected six months after launch.
2. Start with the customer. Describe the improvements for the customer. Outline why this is significant for the customer. How does it differ from competitor solutions?
3. Articulate audacious and clear goals. Articulate clear, measurable results that you have achieved to include financial, operating and market share results.
4. Describe the hurdles you have overcome. Identify the hard results accomplished to achieve success, the important decisions, and the design principles that resulted in success.

4. Bricolage

Bricolage is a way of constructing prototypes using whatever materials are readily available. Bricolage combines practical craftsmanship with a creative and resourceful imagination and an ingenuity that is prepared to use whatever comes to hand in order to make a concept more tangible.

In an entrepreneurial context, bricolage is broadened from physical artefact to process improvement, where it may also consider the available assistance from your wider network. Here, it aligns with lean business and effectual entrepreneurship principles (Sarasvathy, 2008), which have a bias towards making the most of what you have available.

Examples

Figure 6.22 shows how Martin Ryan, designer of the cantilevered horse saddle, re-purposed readily available pieces of equipment to serve as 3D

FIGURE 6.22 Bricolage example.

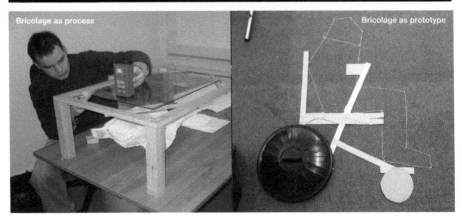

Bricolage as process

Bricolage as prototype

scanners. This was in 2007, when 3D scanners were scarce and expensive. Martin built a catalogue of horse shapes using a room measurement device with addition of glass plate and grid to collect arrays of measurement points which were manually transferred to a CAD package for the final shape to be drawn.

BIBLIOGRAPHY

Bland, David and Osterwalder, Alex (2020). *Testing Business Ideas*. Wiley.
Blank, Steve (2013). *The Four Steps to the Epiphany* (5th edn). K & S Ranch.
Brown, Tim (2009). *Change by Design*. Harper Business.
Buxton, Bill (2007). *Sketching User Experiences*. Focal Press.
Gunther McGrath, Rita and McMillan, Ian (1995, July). Discovery Driven Planning. *Harvard Business Review*.
Klein, Gary (2007). Performing a Project Pre-mortem. *Harvard Business Review*.
Maurya, Ash (2016). *Scaling Lean*. Portfolio Penguin.
Ries, Eric (2011). *The Lean Startup*. Portfolio Penguin.
Sarasvathy, Saras (2008). *Effectuation – Elements of Entrepreneurial Expertise*. Edward Elgar.

Execute

The objective of the Execute phase is to prove revenue-earning capacity.

When you have substantially validated the *desirability* of your preferred concept, along with aspects of the VFS dimensions, you must test the extended value offering and its business model with 'real-life' testing: real users, real scenarios, real paying customers and real competitors. Here, you seek iteratively and rapidly to refine and amend it to be ready for scaling.

The focus of *Execute* is to find a viable business model that can be the vehicle for achieving targeted scale. The initial business model estimate has been substantially de-risked but still retains significant assumptions as success factors. During *Execute*, you tease these out, make quick and significant decisions, pivot where necessary and converge to a scalable solution.

FIGURE 7.1 Chapter 7 outline.

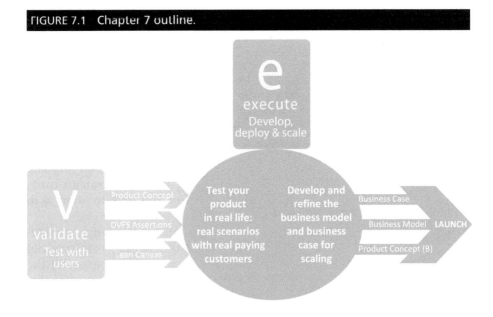

At this time, it is usual that the innovator experiences big pressures, both self-imposed and from the broader organisation, bosses and stakeholders. The innovator and others in the organisation are likely to be impatient. They have been excited by the prospects of the new proposal for some time, and cannot wait to get it quickly to the market.

In this chapter, you will learn how to bring your project from concept to concrete, and make it ready for building to scale.

SO YOU THOUGHT YOU WERE FINISHED?

When is your new innovation product, service or system finished? Probably never! Most likely, you will never reach a time when you can wipe your hands, say that is done, walk away and know that you have done a perfect job. Yet, the time must come when you make a call and turn your attention from trying to perfect the concept towards exploiting the outputs of the effort, time and money that you have invested in the project so far. It is time to put your innovation efforts to work. The overall offering may not be perfect yet, but some parts of it may be tested only in the heat of action.

During the *Validation* phase of the last chapter, we encouraged you to create *assertions* of the key success factors for your project under the headings of *desirability, viability, feasibility* and *strategic suitability*. We gave guidance on how to map and prioritise the assumptions implicit in these factors. We showed you various methods, just a selection from an almost limitless range of possibilities, that you might use to prototype and test the *desirability*-oriented assumptions. In the phrasing of the Lean Startup methodology, the desirability validation efforts have brought you to a state of *problem–solution fit*. You know you have conceived a solution that solves a real problem and brings significant value to your users. It helps them achieve outcomes they need for particular jobs they are trying to complete in their lives.

Additionally, in the last chapter, we noted that you must appropriately investigate the technologies, products and systems associated with the *feasibility* dimension, plot their roadmap, and plan their development. You might well have progressed this development activity by now. For some innovations, technology and product development will be, by far, the longest duration and/or most expensive phase of the project. We did not detail specific methods for technology search, evaluation and development. This is beyond the scope of this book, and is covered by other excellent texts.

Further, we discussed steps you might take to ensure your proposed innovation remains *strategically suitable* for adoption by your firm or by another vehicle that will bring it to market. This involves an ongoing cultivation of support, advice and understanding about the strategic capabilities, directions and priorities within your firm and how to align with them.

Finally, while you made early estimates of your solution's *viability*, based on your developed understanding of the market, you have not yet been able to execute substantial testing to gauge the accuracy of the assumptions implicit in the *assertions*. This is the principal topic of this chapter, and it marks a significant new step along the way to full execution. Now, finally, your proposed solution goes into deployment with everyone expecting that it will add clear and substantial new value to the user, and will be sustainably profitable for your firm. Will the expectations be satisfied?

FOCUS ON VIABILITY AND PRODUCT–MARKET FIT

How many users could get substantial value from your offering? How many will still feel compelled to acquire the new value under the conditions that you will supply it to them? How will you deliver it to them in a way that makes it most distinctive and attractive for them and allows you to build sustainable scale of your business? In other words, is there a viable market and business model?

There are still many assumptions embodied in the core product and business model, particularly in the latter because you have not yet tested it in detail. It is too early for large-scale investments and, before progressing to a full-scale drive, you need to confirm *product–market fit*, which means that the productised solution will be compelling to a sufficient size of market and that the business model to deliver it will be economically sustainable for your firm.

Testing and dealing with the embedded *viability* assumptions has become the main determinant of project success. Building towards *product–market fit* requires continuing the iterative and co-creative interaction with your customers. This is what we have been doing throughout the ARRIVE process from *Research* through *Validate*. However, now the stakes are significantly higher, and the venture must be de-risked. Before going for scale, how will you confirm that your proposed customers – and product with business model – will behave in the way you hope? How will they hear about your product, how will they assess its value, and what will compel them to buy it? How, really, will they use it? Is it the way you had designed and planned for? What will be their overall ownership experience and will they promote you and your product to their friends and colleagues?

Inevitably, your first assertions, being assumptions, will be off the mark in many respects so, to make progress, you are required to tweak, amend, discard or create elements of your business model, or indeed make a major pivot, until the product–market fit is achieved. As an extreme measure, you might even abandon the project and cut your losses. The assessment of customer and competitor behaviours may be confirmed satisfactorily only in the real market, with real product, real customers, real competitors, real performance and real payment.

CONVINCING INVESTORS TO STAY WITH YOU

Despite the pressures to proceed at pace, before you embark on a new phase of activity that demands increased investment levels, it is important you capture what you believe is your business case in a manner that helps you convince colleagues and bosses that you are on to something good, even great, and they should stick with you. You need their support, both moral and financial, as you proceed towards launch.

Every firm has its own processes for investment approval. More often than not, they involve rigidly defined formats that, while maybe necessary, also obscure a clear overview by virtue of their volume and detail. With the wood hidden by the trees, it can be hard for the uninitiated or someone with multiple lines of other responsibilities to get a clear, full picture. Hence, we always recommend supplementing pro-forma spreadsheets and other formal, traditional documents with dashboards that allow you clearly to provide the overview of your proposal and business case, for those executives whose support is essential. The other documents are available for answering detailed enquiries, after they have assimilated the overview. Here is what we use, all three or a combination as may suit the purpose.

Product Concept Board (B)

Not very imaginatively, perhaps, we label this 'B' because it is meant to be an 'inward-facing' item, in contrast to the customer-facing Concept Board 'A' we discussed during concept development in the Ideate chapter. This Concept Board (B) clearly shows to colleagues and bosses what is the essence of the value proposition you intend to offer, along with its origin and outstanding uncertainties. See Figure 7.2.

Business Case Dashboard

Throughout the Validate phase and into this culminating phase of the project you have developed and partially tested your *assertions* surrounding the *viability, feasibility* and *strategic suitability* of your proposal. The Business Case Dashboard packages these in dashboard format for easy overview. See Figure 7.3.

As you start to *Execute* your proposal, the Business Case Dashboard is a key tool in persuading investment for the limited launch phase, described in the next section. As you learn from the experiments of the limited launch, you update its contents and prepare for full-scale launch with an up-to-date Dashboard in which you have an increasing level of confidence.

Lean Canvas

You started working with the Lean Canvas during *Validate* as a tool to tease out elements of the business model through which you would create, deliver and capture value with your proposition. As you learn from the experiments and experience of the limited launch, you will update and fill out its contents and configuration so that you prepare for full launch with a tested and proven business model.

We describe the first two of our three investor-facing canvases (Product Concept Board (B) and Business Case Dashboard) in more detail in the methods section at the end of this chapter. For the Lean Canvas, we refer the reader to one of the many excellent texts on the subject, and we provide reference reading in the bibliography.

FIGURE 7.2 Product Concept Board (B) example.

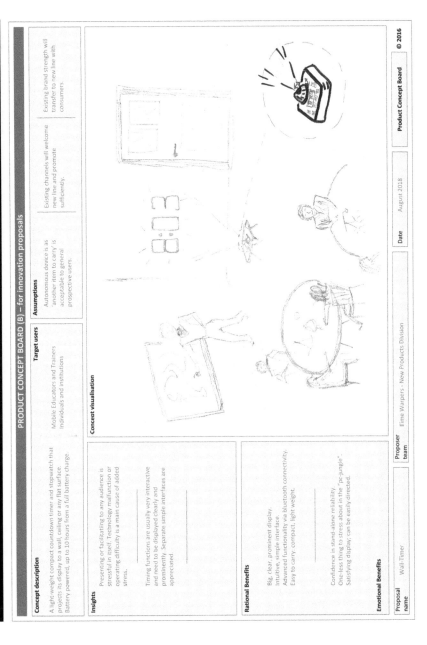

PRODUCT CONCEPT BOARD (B) – for innovation proposals

Concept description

A light-weight compact countdown timer and stopwatch that projects its display to a wall, ceiling or any flat surface. Battery powered, up to 10 hours from a full battery charge.

Target users

Mobile Educators and Trainers
Individuals and institutions

Assumptions

Autonomous device is as 'another item to carry' is acceptable to general prospective users.

Existing channels will welcome new line and promote sufficiently.

Existing brand strength will transfer to new line with consumers.

Insights

Presenting or facilitating to any audience is stressful in itself. Technology malfunction or operating difficulty is a main cause of added stress.

Timing functions are usually very interactive and need to be displayed clearly and prominently. Separate simple interfaces are appreciated.

Concept visualisation

Rational Benefits

Big, clear, prominent display.
Intuitive, simple interface.
Advanced functionality via bluetooth connectivity.
Easy to carry; compact, light weight.

Confidence in stand-alone reliability.
One-less thing to stress about in the "pc-jungle".
Satisfying display, can be easily directed.

Emotional Benefits

| Proposal name | Wall-Timer | Proposer team | Time Warpers - New Products Division | Date | August 2018 | Product Concept Board |

© 2016

FIGURE 7.3 Business Case Dashboard example.

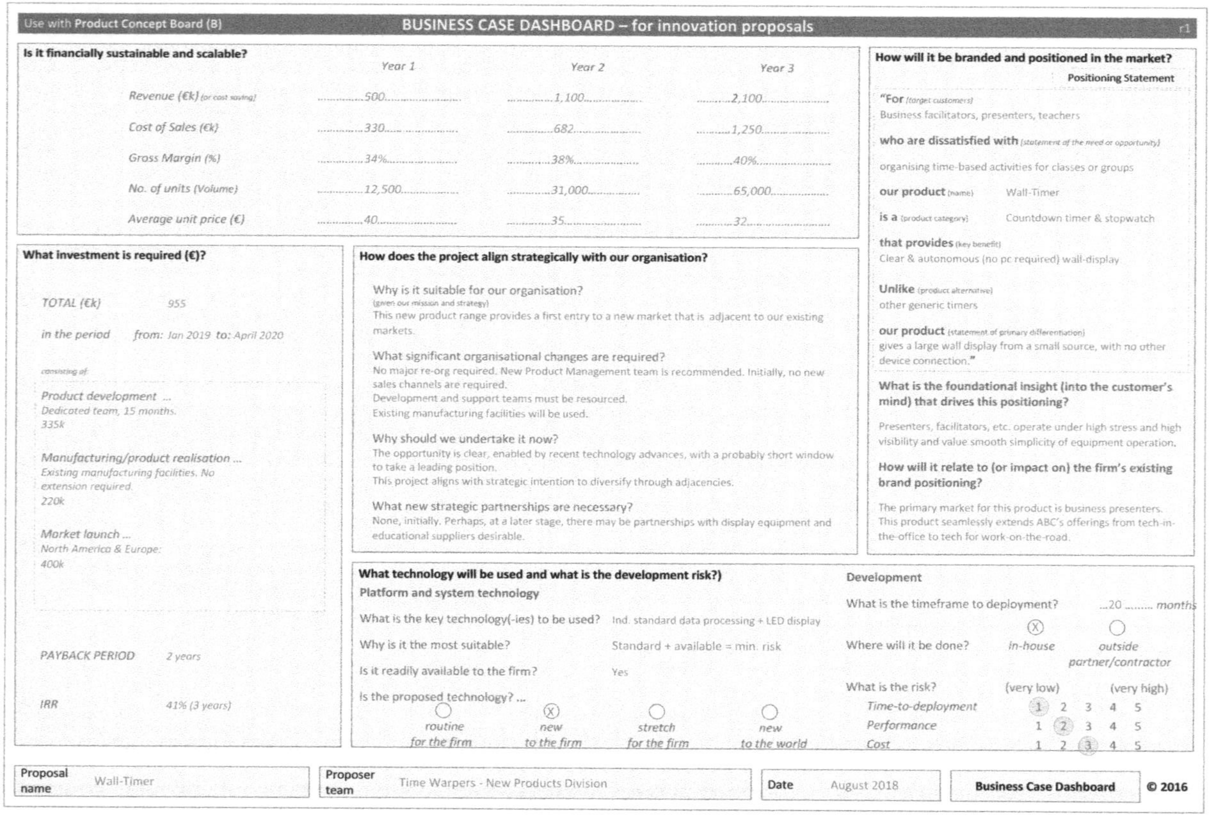

Use with Product Concept Board [B] | **BUSINESS CASE DASHBOARD – for innovation proposals**

Is it financially sustainable and scalable?

	Year 1	Year 2	Year 3
Revenue (€k) (or cost saving)	500	1,100	2,000
Cost of Sales (€k)	330	682	1,250
Gross Margin (%)	34%	38%	40%
No. of units (Volume)	12,500	31,000	65,000
Average unit price (€)	40	35	32

What investment is required (€)?

TOTAL (€k) 955

in the period from: Jan 2019 to: April 2020

comprising of:

Product development ...
Dedicated team, 15 months.
335k

Manufacturing/product realization...
Existing manufacturing facilities. No extension required.
220k

Market launch ...
North America & Europe.
400k

PAYBACK PERIOD 2 years

IRR 43% (3 years)

How does the project align strategically with our organisation?

Why is it suitable for our organisation?
(given our mission and strategy)
This new product range provides a first entry to a new market that is adjacent to our existing markets.

What significant organisational changes are required?
No major re-org required. New Product Management team is recommended. Initially, no new sales channels are required. Development and support teams must be resourced. Existing manufacturing facilities will be used.

Why should we undertake it now?
The opportunity is clear, enabled by recent technology advances, with a probably short window to take a leading position. This project aligns with strategic intention to diversify through adjacencies.

What new strategic partnerships are necessary?
None, initially. Perhaps, at a later stage, there may be partnerships with display equipment and educational suppliers desirable.

What technology will be used and what is the development risk?

Platform and system technology

What is the key technology(-ies) to be used? Ind. standard data processing + LED display

Why is it the most suitable? Standard + available = min risk

Is it readily available to the firm? Yes

Is the proposed technology? ...

routine	new	stretch	new
for the firm	to the firm	for the firm	to the world

How will it be branded and positioned in the market?

Positioning Statement

"For (target customers)
Business facilitators, presenters, teachers

who are dissatisfied with (statement of the need or opportunity)
organising time-based activities for classes or groups

our product (name) Wall-Timer

is a (product category) Countdown timer & stopwatch

that provides (key benefit)
Clear & autonomous (no pc required) wall-display

Unlike (product alternatives)
other generic timers

our product (statement of primary differentiation)
gives a large wall display from a small source, with no other device connection.'"

What is the foundational insight (into the customer's mind) that drives this positioning?
Presenters, facilitators, etc. operate under high stress and high visibility and value smooth simplicity of equipment operation.

How will it relate to (or impact on) the firm's existing brand positioning?
The primary market for this product is business presenters. This product seamlessly extends ABC's offerings from tech-in-the-office to tech for work-on-the-road.

Development

What is the timeframe to deployment? 20 .... months

Where will it be done? In-house / outside partner/contractor

What is the risk?

	(very low)				(very high)
Time-to-deployment	1	2	3	4	5
Performance	1	2	3	4	5
Cost	1	2	3	4	5

Proposal name Wall-Timer

Proposer team Time Warpers - New Products Division

Date August 2018

Business Case Dashboard © 2016

LIMITED LAUNCH

A limited launch is employed in order to confront risk with a tightly targeted and time-bounded first step into the market.

Eventually, you must bite the bullet in order to get valid experience of real market conditions, but you cannot risk a full-scale launch with still many uncertainties. The solution is a *limited launch*, a limited range, perhaps low-key and probably highly targeted launch. Liedtka and Ogilvie (2011) call this a *Learning Launch*. The real product is doing a real job for a real customer but has a minimal set of features. This describes a minimum viable product (MVP), as we shall discuss below. Depending on your business type, you may select to execute your MVP's limited launch to a few preferred, friendly customers or you may opt for narrowly bounded geographical or sectoral segments. Alternatively, for a custom problem solution, you may deploy and operate your minimal custom solution in the targeted use environment, with some operational and environmental restrictions.

Ideally, you will impose the full, intended market price. This is the best scenario from a realistic testing perspective. However, remember that the objective of this *Execute* phase is not to earn revenue but to prove revenue-earning capacity, and margin capacity, and thereby reduce risks of scaling to high volumes. The paradigm is clear. When your customer experiences a realistic market situation, she also expects, and will demand, good value in return for payment. There will not be any doubt about how to interpret her response.

In every case, the consequences of failure at this stage are always higher than during concept development. Usually there have been substantial manufacturing or solution realisation costs incurred, despite your attempts to minimise these at the start. In the heavy hardware, complex systems and pharmaceutical industries, for example, these can be extremely high for new products even for limited use. Also, there may be substantial market launch promotional campaigns and channel priming costs. Additionally, reputational risks come to the fore for established firms – slightly less so for startups – and must be considered strategically. Until now, engagement with customers and

FIGURE 7.4   Risk, hence assumption, profile of the ARRIVE process versus traditional innovation processes.

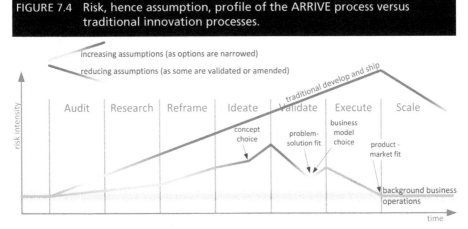

At times of critical choice, progress requires decisions and the inherent assumptions crystallise as risk.

users has been under the banner of product development, with no specific short-term promise of performance, and has depended on the freely given goodwill of the customer. From now on, your engagement with customers and users carries a clear promise of performance and satisfaction in their operational lives – after all, the customer is paying for this – and the pressures can become intense!

MIND THE INTELLECTUAL PROPERTY

Also, at this stage intellectual property protection issues may come to the fore, for which you must be prepared. We do not go into this in detail here; there are many other texts and sources of advice. Our comments here are a reminder not to forget this in the midst of the heady rush towards market presence. You must be diligent in protecting patentable and other protectable intellectual property components of your innovation before you bring them into the public domain.

Up to this stage, non-disclosure arrangements have easily provided umbrella cover for confidentiality in your engagements with users. Before going public, however, you must perform a thorough review of how you propose to manage the intellectual property assets incorporated in your innovation. For which aspects should you consider patent, copyright, trademark registration? Which aspects do you need to protect by keeping them hidden or 'trade secret'? For which territories? Remember, international patent registration and maintenance costs can be a heavy time and money burden. Is it worth it? Are there third-party intellectual property assets that you plan to utilise? Have you negotiated the clear authority and conditions for their use?

If your project is for a larger firm, it is likely that there will be in-house IP management supports for you to draw on. If these are not available, you must seek out external advice so that you may have a robust protection in place, and that you will not be blindsided later by copycat competitive encroachment.

A RACE AGAINST RESOURCES TO PRODUCT–MARKET FIT

Liedtka and Ogilvie (2011) advise, and we concur, that this phase must be time-bounded and budget-bounded and clearly planned as such. Comparably, in the Lean Startup methodology for new firms, Steve Blank often makes the point that a startup is partly defined by having limited resources, and it is the race against depleting budget that compels the entrepreneur to extraordinary effort and ingenuity in the search for product–market fit. In the large corporate case, budget controls for a project are often more fluid, and there can be a danger that the momentum of the project will push it into the *Execute* phase without clear plans or go/no-go criteria. This could be fatal, because *Execute* has a clear purpose, beyond just a gateway. If you cannot prove a business model is suitable for committing investment to support full-scale deployment, then the project should be abandoned or substantially pivoted. Because your search involves substantial costs of time, money, human resources and reputation, these are not issues that you can let slide indefinitely and a clearly bounded resource availability provides the limit.

On the other hand, in the *Execute* phase the product intersects with the market in a real, nuts-and-bolts way. Hence, the project needs proper resourcing, with developers and others to create the first product or MVP and to iterate improvements continuously. In addition, marketing and product management resources will be heavily involved as well as manufacturing and project management. With a demanding objective to remove uncertainty regarding how, or even if, the innovation can be successfully deployed at scale, you must avoid the assumption that a natural, smooth transition to full-scale launch is automatic. All hands are required on deck!

Note regarding terminology

Until now we tended to refer more, though not exclusively, to 'users' and 'low-fi' prototypes. Now, we focus on the business model, getting product–market fit, saleability and scalability. Therefore, from here on, we prefer to use the terms 'customers', indicating paying customers, and MVP, indicating real performing product.

ASIDE: COMPARING *VALIDATE* AND *EXECUTE*

Often, we have seen confusion as to what are the differences between the *Validate* and *Execute* phases. True, both phases continue a pattern of intensive co-creation and iterating prototype evaluation with customers and users. Specifically, *Validate* and *Execute* are the phases where assumptions about concept and business model, respectively, are tested.

- In *Validate*, your learning is directed towards core *concept development and testing*. In *Execute*, your learning is for wider *business model development*, as well as concept *confirmation and tweaking*.
- In *Validate*, value flow is mostly that the customer provides value to you, which feeds your concept development effort. In *Execute*, you provide value to the customer, which supports her process or job to be done. More simply: In *Validate* the customer does not pay; in *Execute* the customer pays.
- In *Validate*, we work with rapid low-fi prototypes to aid learning about the core concept or about individual features. In *Execute* we produce MVPs to aid learning about a complete, featured product and business model.

LEAN STARTUP

It is interesting to note that the formal expression of a methodology for deploying products into the market iteratively and co-creatively is a recent development. It has been made prominent and popular in the 21st century, by the Lean Startup movement. In 2003 Steve Blank began publicising his ideas for an entrepreneurial *Customer Development* process in *The Four Steps to the Epiphany* (Blank, 2013) as a complement to the then-dominant product development process for start-up firms. This was further developed by Eric Ries (2011) in *The Lean Startup*, and Ash Maurya (2010, 2016) in *Running Lean* and subsequently *Scaling Lean*. In parallel,

FIGURE 7.5 Lean Startup contributions from Steve Blank, Eric Ries, Alexander Osterwalder and Ash Maurya.

A.

Four Steps to the Epiphany (1st edition: 2003)
by
Steve Blank

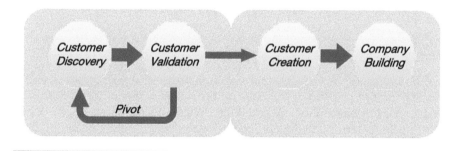

B.

Business Model Generation (1st edition: 2010)
by
Alexander Osterwalder & Yves Pigneur

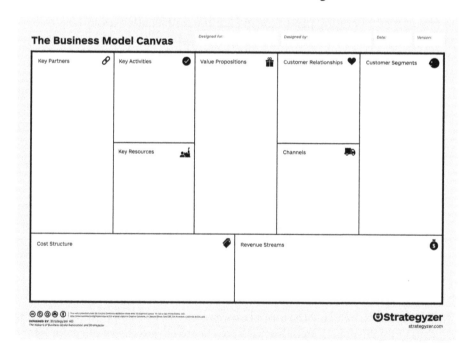

(Continued)

FIGURE 7.5 *(Continued)*

C.

The Lean Startup (1ˢᵗ edition 2011)
by
Eric Ries

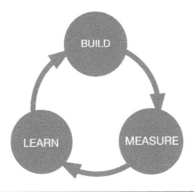

D.

Running Lean (1ˢᵗ edition 2011)
by
Ash Maurya

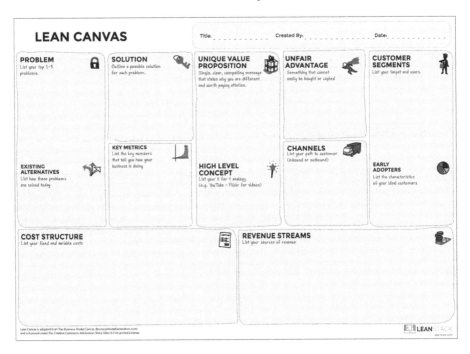

Alexander Osterwalder and Yves Pigneur developed their approach to business model innovation, *Business Model Generation* (Osterwalder and Pigneur, 2010) and in particular popularised the Business Model Canvas. Later, Ash Maurya adapted the Business Model Canvas to create the Lean Canvas as a more actionable version of the tool for driving startup performance. Both the Business Model Canvas and the Lean Canvas have become the foundational tools for the Lean Startup process through the last decade. In this book, we lean towards the Lean Canvas, which we introduced in the previous chapter, *Validate*.

The Lean Startup process takes as its starting point an estimated value proposition and business model as represented on the Business Model Canvas or Lean Canvas. According to Steve Blank, this initial estimate "comes from the founder's vision". The Lean Startup is not concerned where the founder's vision may have come from.

The distinctive jewel of design thinking and design innovation is that our value proposition and business model are the result of a methodical, creative and empathic process of research and concept development, as we have laid out in the ARRIVE framework. In the *Validate* phase we have gained a good degree of assurance about the *Desirability*, or value proposition, part of the proposal.

This is why we paid so much attention to engaging with users throughout the ARRIVE process and in particular to validating our desirability assumptions in the previous chapter. The constant interaction with users, the explicit identification and prioritisation of assumptions, the careful and creative validation of those (*desirability*) assumptions to the best extent possible were all for the purpose of achieving problem–solution fit and de-risking the decision to move on to execution and scaling. During *Execute*, you repeat a similar process of assumption testing and follow-on adaptation of the business model in order to achieve product–market fit, following the rubric of Lean Startup.

TAKING THE PLUNGE

There is some consolation in knowing that even paid-up customers can be forgiving when encountering something radically new and possibly transformative to their lives. Do not forget, early cars of the last century did not have windscreen wipers! Generally, customers do not demand perfection from the start. They may even not recognise exactly what perfection entails, given the novelty of your offering, and they may sense you cannot give it at this stage. But, they will certainly expect close attention and dedicated support response that bridges the gap as rapidly as possible between what they need and what you have supplied.

The stages to be observed when starting to launch your product in the market are:

1. Determine your first customer segment/user type/sector/territory (as appropriate to your project) that you will target for development.
2. Estimate your first product offering and appropriate business model that would bring your solution to scale in the market.

3. List your key outstanding assumptions. These are assumptions that you intend to test before committing to full-scale deployment.
4. Set your metrics and defining levels for success in an Execute Test Plan. Be sure you identify clear, meaningful, measurable and actionable customer development metrics.
5. Stay close to your customers, as individuals and as cohorts. Be voracious in observation, gathering feedback and learning how to achieve or surpass the metrics you have set.
6. Stay alert and fully resourced to iteratively tweak and pivot the value proposition and business model until a valid and scalable business model is identified, which defines a product–market fit.

Let us elaborate on these points.

1. *Determine your first customer.*
   If your solution is a business-to-business capital expenditure product, then ideally you should choose a few, or more, favoured customers for your first installations. It is good if one or some of these are the same people who engaged with you already at earlier stages, through research and early concept validation, because they will be familiar with the project and what it is trying to do and perhaps be more tolerant of early deficiencies. Maybe, they will even feel a sense of ownership. However, be sure also to bring some new customers into the game so that your learning and understanding will not be limited or distorted through familiarity. As you work through these early installations, you will learn much about what works, what does not work and where improvements are needed.
   If your solution is a system sold through dealers or value-adding resellers, for example, you should choose one or two, who have resources and inclination to be cooperative, and work through multiple sale and installation cycles with them.
   If you have a consumer or internet-based product, choose a small sector that sufficiently represents the larger market, while not being so large as to overstrain your supply capacity or destroy your reputation in the larger market, in the event of one or more significant setbacks in your testing.
   Should these initial customers pay? Yes, because you need to establish beyond doubt that you are providing real value. If you provide sufficient value, then the customer will happily pay proportionately. If you provide a free-of-charge offering, you will not be able to determine its value to the customer or the supply chain with sufficient clarity.
2. *Determine your first product offering and business model.*
   Execute is the final and essential preparation for a full-scale launch, and during it the primary objective is to figure out how to exploit your core product with a scalable business model in a viable market. What is the best business model that will encourage your target users to become paying customers? Represent your best business model estimate by continuing to update the Lean Canvas introduced in the previous chapter.

The proliferating studies in behavioural psychology and behavioural economics over the last two decades have shown clearly that predicting human behaviour in unfamiliar circumstances is fraught with error. Therefore, any experimental attempt to support prediction must be close to the projected 'real' conditions in all 'external' aspects of the business model. Here, this means that you may rely on your customers' response behaviours towards your product only to the extent that their product experience approaches what appears real, from the customers' perspectives. The same applies for other external stakeholders with other aspects of the business model. This is important: in order to check your assumptions, your product and business model must give a realistic experience to the customer and external partners. It does not matter how it is produced or delivered, or what it costs you to produce each unit, for the moment, so long as it is received and perceived by the customer and partners in the same way as it will be in full-scale supply.

Following from this, the areas that require particular consideration and judgement are around the feature set and options of your product, and the externally facing aspects of your business model. Probably, your concept involves a vision for a fully featured, extendable product family, including perhaps a range of purchase options and a roadmap for development. However, at the start you need your product to appear real to the customer; hence, it cannot be extensively featured in view of the strict time and financial constraints that you must set. The key question is this: What is the minimum set of features that are necessary in order to (i) deliver the core value proposition to the customer, and (ii) provoke real and relevant responses that you may rely on to guide you towards a final offering and business model?

Generally, a new product must have a core functionality that is compelling on its own. Ancillary features that will be 'nice' are not essential. Remember early cars without windscreen wipers? Or, recall early personal computers without monitors and connection cables! Airbnb started with shared spaces before extending to all types of accommodation! You must seek a way to represent the core functionality clearly and fully to the customer, despite perhaps missing other peripheral features, in order to evoke and observe a truly useful and relevant response. Similarly, any business model elements that rely on external partners or suppliers must be thoroughly evaluated in conditions as near as possible to future reality. This is a minimum viable product, or MVP. It demands close understanding of the customer, carefully considered judgement of the core value proposition you offer, and great ingenuity to create a satisfying MVP at minimum cost. This perspective is closely aligned with the approach of the Lean Startup movement. Eric Ries (2011) says:

> [the MVP] is the fastest way to get through the Build-Measure-Learn loop with the minimum amount of effort. Contrary to traditional product development, which usually involves a long, thoughtful incubation period and strives for product perfection, the goal of the MVP is to begin the process of learning, not

end it. Unlike a prototype or concept test, an MVP is designed not just to answer product design or technical questions. Its goal is to test fundamental business hypotheses.

3. *List your outstanding assumptions.*
Re-visit all your assumptions. Probably, you have by this stage satisfied yourself with regard to most of the *desirability, feasibility* or *strategic suitability* assumptions, but some of these may remain. What are the assumptions related to *viability* (the market and the business model) that remain as key success factors for the project? Use the *DVFS Assertions*, and methods described in the last chapter, to assist your thinking. Prioritise the assumptions according to importance and uncertainty.

4. *Plan how you will experiment to test your assumptions. Set your metrics that define success.*
You must carefully plan how you are going to test each assumption. As we said, it involves a considerable amount of ingenuity to get the most learning bang from a limited number of bucks.

For each test, define the measure that will tell you whether the test has been passed. If you do not objectify the metrics at this stage, it is probable your emotional attachment to the project will cause you to be too forgiving of failures or 'almost' successes when they inevitably occur.

With clear and dispassionately considered metrics of success, resolve not to proceed with the project until each metric has been surpassed – or you can rationally justify an alternative metric or plan. For example, for an assumed price point of €1,500, your metric might be that you can get at least 20 paying customers in your first three months, and 50 paying customers in your second three months. If you fail to achieve your predicted numbers by a substantial gap, then you must seriously question your understanding of the value proposition you offer, its value to the target customers you have chosen, the efficacy of the delivery model you have used, or all three. Something is wrong, and it may be fatal! You have to triage it, identify the problem and test again, urgently.

We consider in more detail in the next section the topic of planning your experiments.

5. *Stay close to your customers.*
It is not enough to take actions like promoting, offering, selling and installing. Even doing these well and with some overall success provides only surface level information. You have to keep working on it, dig deeply and stay alert to understand which parts of your overall offering and business model work well and what might be improved or eliminated. To achieve this, you must stay close to your customers all the way through the *Execute* phase, as individuals and as cohorts. Your focus is to learn from them what works, through qualitative enquiry such as observation and interviews as well as through quantitative assessments such as rates of customer acquisition and conversion, repeat business rates, etc. using whatever may be appropriate for your particular business.

When you make changes to your business model, be sure you set new metrics and success criteria as appropriate and in a manner that enables you reliably to attribute changes in performance to the changed business model.

6. *Stay alert, fully resourced and agile.*

If something is clearly not working as productively as you expected, change it! Search thoroughly to find these flaw conditions and try hard to prove your plan and expectations wrong by keeping a constantly sceptical mindset. Remember, after full launch the reality of the market will find remaining flaws in super quick time, when it will likely be too late for you to do anything about it without major cost and upset. Be ready continuously and iteratively to tweak and pivot the value proposition and business model in response to your observations.

However, do not be too hasty! Some actions have natural response delays and you must wait for these to take their natural course before making judgement. For example, if you give a business customer a quote for €2,500 to supply a piece of equipment, it might take some time for this to be considered or signed off by a senior manager or by Procurement. It would not do for you to lose patience, assume there is a problem after a week and drop your price to €2,000 to encourage the sale. As another example, viral spreading of messaging or collaborative working applications usually requires a tipping point of user numbers to be attained before really strong growth happens. Before tipping point level, growth is slow and fuelled by innovators and early adopters, while mainstream growth will be seen only after adoption has tipped into the early majority. On the other hand, as soon as you can be sure that an action or element of your model is not working, you must work out what an alternative may be and set about implementing it as a matter of urgency.

Execute requires a dedicated multidisciplinary 'SWAT' team, primed to manage and promptly respond to all of the mini crises that will arise. Strategic marketing, product management and development engineering must work hand in hand more closely than ever before to absorb the rapid learning and deliver the appropriate rapid responses. As well as long-standing team members, it is advisable to include some newcomers to the team at this stage, who will be better able to take an independent and unbiased view of data and progress. Constant, diligent monitoring is key. Daily stand-up meetings that review recent data and discuss planned action responses are essential to keep the whole team up to speed and working in unison in the midst of hectic flows of product issues, actions, communications, field data and new decisions.

PLANNING YOUR TESTS TO BUILD YOUR BUSINESS MODEL

Figure 7.6 shows a short extract from the Priority Assumptions List used by Martin Ryan in the development of a revolutionary new cantilevered saddle. The team assembled this list after first-stage desirability testing, which encouraged them that they were converging towards a good problem–solution fit, after months of prototype iteration and testing with horse-riders from different sectors of the equine sports industry.

As it entered the *Execute* phase for the new saddle, the team prepared a *Limited Launch Test Plan* to determine the reliability of assumptions and flesh

FIGURE 7.6   Priority Assumptions List (extract) for the Cantilevered Saddle by Martin Ryan.

Priority Assumptions List
Desirability | Viability | Feasibility | Strategic Suitability

Desirability

1. The saddle's characteristics provide particularly great benefits for the 'eventing' competitors, by its modular design and suspension characteristics.

2. The cantilevered suspension tree will afford a more comfortable experience for the horse and rider and still support sufficient connection between both to deliver performance.

3. The saddle is suitable for the highest level of jumping competition.

Viability

1. The best sectoral target for launch is 'eventing', which has good pick-up rate prospects and price-per-performance potential. By tradition, eventers are open minded to new technology and solutions.

2. ISB Inc. will be our exclusive partner for North America for import, distribution and support.

3. Expensive promotion will not be necessary. The novelty of the design will continue to attract media interest and drive a solid customer base in the first few years

4. The recurring sales revenue potential afforded by its modular design will attract resellers to stock the saddle and sell for lower than standard margins.

5. The saddle will lend itself to online sales, even without trial. Testimonials and promotion video will satisfy the concerns and questions for early online customers.

Feasibility

1. Our subcontract manufacturer and assembly facilities can ramp up production volume at short notice.

2. There is no risk of supply line blockages of the key composite material for the saddle frame.

Strategic Suitability

1. Our main investors are still fully behind growing this project to a scaled up product line, with new investment support as required.

2. Our key staff (still) have appetite for the intensive effort over the next 18-24 months (to reach stability).

out flaws and weaknesses. This plan resulted from many hours' consideration and discussion, and some contested arguments. It interacted closely with the choice of initial test customers, MVP and business model. For example, it assumed a particular business partner as a key element of the business model. If the collaboration of this assumed key partner could not be verified satisfactorily, it would be prudent to change to another partner. The level of certainty required at this stage requires testability of all main criteria. Table 7.1 shows an extract from the Cantilevered Saddle's *Execute Test Plan*.

Assumption V1 (*Viability*) of the Cantilevered Saddle's Test Plan was that *Eventing* is the most attractive segment. To test this exactly requires testing many other sectors and comparing results between sectors. Of course, there is a practical limit to the amount of testing that can be done and, in this case, Martin Ryan's team has opted more pragmatically to confirm that the Eventing sector is a good segment, based on the metrics chosen. The team relied on

Table 7.1 Execute Test Plan (extract) for the Cantilevered Saddle by Martin Ryan.

Cantilevered Saddle – Execute Test Plan		
Assumption	Hypothesis to test	Experiment	Success measure
DESIRABILITY			
D1. The saddle's characteristics provide great benefits for the 'eventing' sector by its modular design and suspension characteristics.	The modular design appeals to semi-professional eventing riders, as they need different types of saddles for different parts of the event. The modularity offers a more cost-effective solution to multiple saddles.	Demonstrate the saddle to a variety of horse disciplines. Assess the reaction of eventers compared to other disciplines.	Eventers must show a tangible excitement concerning the saddle's prospects and strong intentions, with some commitments, to purchase the saddle.
D2. The cantilevered suspension tree will afford a more comfortable experience for the horse and rider and still support sufficient connection between both to deliver performance.	The cantilevered saddle is more comfortable for horse and rider.	Test with different types of horse and rider. Seek testimonials from riders. Seek testimonials from specialists such as horse physiotherapists and veterinary practitioners.	>50% of test riders should report improved comfort for themselves and perceive improve comfort for the horse. + <20% have serious doubts or faults.
D3. The saddle is suitable for the highest level of jumping competition.	Top-level international riders will be willing to use the saddle in Grand Prix competition.	Conduct trials with the top riders in the army equitation school and other international grade riders.	Good level of general interest and at least one professional rider adopts a firm intention to use the saddle in competition in the short term (three months).

VIABILITY

V1. The best sectoral launch target is 'eventing', which has greatest pick-up rate prospects and price per performance potential.	(a) Our target retail price is reasonable for this market. (b) There is a fast rate of adoption of novelties in this market.	Arrange for >10 professional and hobby event riders to test the saddle over a period of multiple weeks.	(a) >70% of riders who have trialled the saddle confirm the retail price they are willing to pay is at a level greater than our target retail price (using a suitable, non-leading testing methodology). (b) >50% confirm they will consider buying the saddle within six months.
V2. ISB Inc. will be our exclusive partner for North America for import, distribution and support.	ISB will aggressively promote the cantilevered saddle as their main product line for eventing. ISB will accept our conditions of business.	Establish (or ramp up) strategic contacts and negotiations to flesh out the appetite and ISB's approach to our new business.	Heads of Agreement signed within six weeks Commitment to invest in stock and Sales staff training at a rapid ramp-up rate.
V3. Expensive promotion will not be necessary. The novelty of the design will continue to attract media interest and drive a solid customer base in the first few years.	Customer acquisition will be achieved most by referrals through social media and personal contacts.	Open a Facebook site, with a blog and heavily promoted. Seed this with technical descriptions, videos, testimonial, and links to other equestrian sites/blogs.	A high rate of viewers, blog participants and referrals with high 'likes' and positive comments.
V4. The recurring sales revenue potential due to its modular design will attract resellers to stock the saddle and sell the basic saddle with lower than standard margins.	Established saddle sellers will be keen to stock and sell the saddle.	Meet with >5 established saddle sellers/stockists in each of Ireland, Germany and Australia.	At least two in each territory must be keen to stock the cantilevered saddle.
V5. The saddle will lend itself to online sales without rider testing. Testimonials and a promotion video will satisfy the concerns and questions for early online customers.	International customers will order the saddle from an online shop.	Offer the saddle for sale online, with intensive four-week promotional campaign (Facebook and Google Ads) and launch at the Dublin horse show. Show pics and testimonial from two well-known professional riders.	Secure ten serious enquiries and three sales within four weeks.

(Continued)

Table 7.1 Execute Test Plan (extract) for the Cantilevered Saddle by Martin Ryan.   *(Continued)*

Cantilevered Saddle – Execute Test Plan

Assumption	Hypothesis to test	Experiment	Success measure
		FEASIBILITY	
F1. Subcontract manufacturer and assembly suppliers can ramp up volume quickly.	(a) If our demand increases, the manufacturer and assembly suppliers will be able to respond to extra demand within one month. (b) Alternative/additional suppliers are available.	Conduct suppliers audit. Space, machine, labour capacity. Critical materials – global market availability. Supplier history of flexibility with other customers. Conduct supplier market audit with preliminary enquiries.	No major flaw found. At least two alternatives for all existing key suppliers is currently available.
F2. There is no risk of supply line blockages of the key composite material for the saddle frame.	There are alternative sources for the material and no indication of world shortage.	Map the supply chain and identify risk characteristics of each stage.	No single point of failure without alternative.
		STRATEGIC SUITABILITY	
S1. Our key investors are still behind us.	When presented with successful '*Execute*' phase outcomes, investors will be willing to provide the needed support.	Prepare an investment plan based on successful Execute outcomes. Ask them to commit!	At least 50 per cent of investors (if n>6) or oversubscription to 150 per cent promised (whichever is higher).
S2. Our key staff (still) have appetite for the intensive effort over the next 18–24 months.	(a) With the clear understanding of the road ahead, no one is considering bailing out.	Conduct interviews – deep 'n' meaningful conversations with each key staff member. Understand their stresses and expectations. Make a judgement for each.	No key staff member is likely to leave, at least not without warning as to growing stresses, i.e. a good interpersonal relationship and understanding exists with all.

its own intuitive sense of the broader market to accept that eventing was 'good enough', leaving aside the moot question of whether it is 'the best'.

Some authors and consultants propose a strict evaluation of experiment outcome against pre-set metrics. If the metric is not satisfied by the experiment outcome, then the business model is adjusted, a new experiment is run, and this is repeated until a 'pass' is achieved. This is a good principle, but in practice sometimes it is not so clear-cut and straightforward. Extremes of performance are easy to act on; but, results that hover close to the 'pass' level need finer judgement. Perhaps the narrow 'fail' is easily explainable. Perhaps, in hindsight, the metric level was chosen to be unnecessarily ambitious and a new level is now justifiable based on new learning. In our view, narrow passes of pre-set metric levels, or failures to pass, in one or two experiments may not give cause for major disruption to an *Execute* plan. However, narrow failures (or narrow passes) across many experiments surely indicate a project struggling to succeed and with little reserves for contingency against a future unexpected occurrence. This deserves real pause for consideration.

FROM DESIGN INNOVATION TO LEAN STARTUP

You must carefully design, execute and evaluate each element of the test plan. You must rigorously consider the nature of your minimum viable product. What is the core value proposition? What must you include and what may you leave aside for now? The MVP is the tool that brings the core value proposition to the subject (user), in such a way that you get a response you can evaluate. Additionally, you must define clearly the metrics that will capture that response unambiguously and reliably, and be sure you are capable of measuring those metrics. This is the domain of Lean Startup.

Lean Startup is a natural sequel to Design Innovation in business innovation. Both hold sacred a continuing intimate engagement with the user. Design Innovation explores the user's problem space and conjures multiple solution concepts. It evaluates the options and validates the chosen concept through repeated experimentation with users. Its natural focus is to achieve problem–solution fit. Lean Startup continues this process by validating the business model experimentally with users in a competitive market context, with a focus on product–market fit.

We refer you here to the excellent books on Lean Startup referenced in this chapter, which you can use for further detailed guidance on these areas. Remember, as it has been throughout the concept validation phase of your project, your continuing task is further de-risking of your innovation proposal by experimentally reducing the uncertainty embedded in your assumptions.

FROM RUNWAY TO TAKE-OFF

The *Execute* phase of design innovation may remind you of a radically new aircraft design on the test runway for its first test flight. Simulations have been done, and it should fly. You are the chief engineer and test pilot. Runway length is limited and there are some headwinds. But, you have designed the plane for these conditions so you set an initial course, gather speed, check wind direction, increase momentum, continually correct course and hope that you do not run out of tarmac before reaching take-off velocity.

In the face of excessive head winds, aerodynamic imbalance or component malfunction, you may be able to compensate safely and carry on, or you may abort your take-off for this attempt. Cancellation may be embarrassing, but not avoiding an imminent catastrophe would be worse. If you postpone your initial launch attempt, it does not mean you must abandon the project. It means returning to the hangar, reviewing your previous flight plan, understanding why it did not succeed this time and then adjusting for another go, if still you can summon the energy and resources to do so. This is a major pivot.

Of course, in most cases, especially when the preparatory work has been well done, there is enough adjustment in the aircraft control system to accommodate typical variations in runway conditions and successful take-off results at the first or second attempt. This is business model tweaking and evolution.

Aerodynamic balance of your project represents a balanced performance achievement across all four factors of innovation that we have discussed at length in this book – *Desirability, Viability, Feasibility* and *Strategic Suitability*. When you repeatedly meet and surpass the metrics you have set, then you know you have reached take-off velocity.

Once take-off has been achieved you must set a destination, chart a course and keep adjusting for changing environmental conditions on the journey.

FIGURE 7.7   A balanced performance across the four factors is the goal.

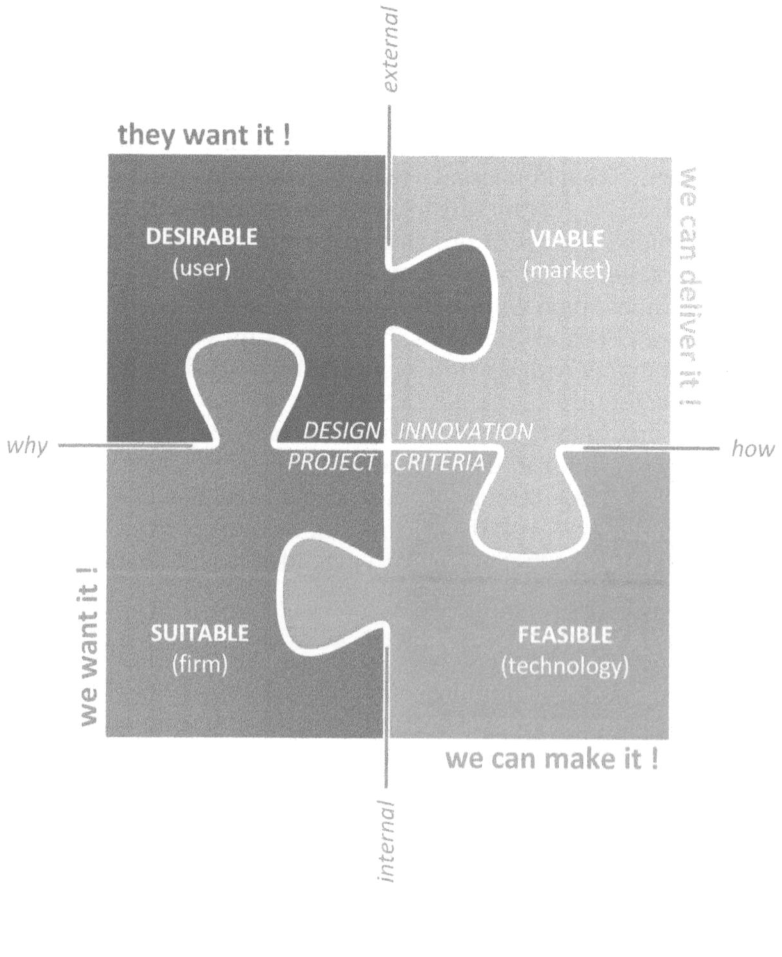

These more classic issues of business development strategy, strategic marketing, roadmap planning, market roll-out and scaling are covered exhaustively in many business texts and we refer you to these for more detail. The outputs of the *Execute* phase leave you with adequate evidence to plot your future and justify your business case with confidence.

METHODS IN THIS CHAPTER

The key activity of the *Execute* phase is devising a test plan that evaluates, refines and iterates both product concept and business model. Inevitable earlier uncertainties are eliminated or at least limited. The proposed scaling is de-risked.

In the methods section of this chapter, we provide instructions on preparing the following support documents:

1. *Product Concept Board (B)*
2. *Business Case Dashboard*
3. *Execute Test Plan*

Execute | methods

PRODUCT CONCEPT BOARD (B)

Determine best estimates of a business proposal across all DVFS dimensions.

BUSINESS CASE DASHBOARD

Create an overview of the business case surrounding the business proposal to support and complement the Product Concept Board (B)

EXECUTE TEST PLAN

Rigorous marketplace testing of the Execute's Limited Launch requires careful and astute planning

Prepare the innovation proposal for a limited launch deployment, where you prove the business model's revenue-earning capacity and scaling potential, before full-scale launch.

"I only made this letter longer because I had not the time to make it shorter."

Blaise Pascal, Lettres Provinciales, 1656

The word concept means different things to different folks in different situations.

At earlier stages of our innovation process, we use the word to describe a high-level representation of a proposed solution, with a purpose to bring it to the intended user and give a good understanding of what is proposed in order to get feedback and advice to help improve it.

Now, you turn your attention inwards to your colleagues and bosses, as you must convince them that your developed proposal is worthy of investment.

This (B) version of concept explains the high-level value proposition and to whom it is being offered. It explains the basis (insights) from which it has been developed and highlights the key benefits the user will experience. Most importantly, it signals the principal outstanding assumptions or uncertainties surrounding its projected adoption and success.

Above all, the Concept Board (B) provides a visual overview of the proposition, whether it is a physical product, service or system. When displayed as a poster, the viewer can absorb a comprehensive overview of why this concept should be attractive to the proposed users.

Step by step

1. Gather your thoughts and your data
Is it time to call a halt to the intense round of validation activities? Gather your senior team members, for a review and reflection:

- Have we reached a stage of saturation, i.e. diminishing returns on efforts?
- Have we achieved our goal of identifying a super offering that fulfils our users' desired outcomes?
- Are we prepared to make the judgement that our proposal looks set to meet the viability, feasibly and strategic suitability criteria, even though some of the business model and other issues may not yet be fully evaluated?

2. Build the Concept Board (B) elements
If the mood is positive about going ahead, start to build your Concept Board (B) elements. This takes time and merits undiluted focus by all. Your team knows the proposition intimately so your difficulty is similar to the quote by Blaise Pascal at the top of this page. Pick out only the principal highlights that form the essence of the proposal. Do not include anything that is not essential information for understanding.

Prepare this at first on a whiteboard where you can make changes easily. Replicate the template shown on the next page. Give particular attention to the visualisation. It should avoid detail, while showing context, key features and users. It may take the form of product sketch, service storyboard, system flow chart or anything that will clarify the essential concept for the viewer.

3. **Construct the Concept Board (B)**

When the team is agreed the concept is well represented, transfer it to a good-sized template as a poster (e.g. A1 size, 2 ft × 3 ft or similar). Do not overfill the template; keep it easy to read and absorb.

PRODUCT CONCEPT BOARD (B) template 2/4

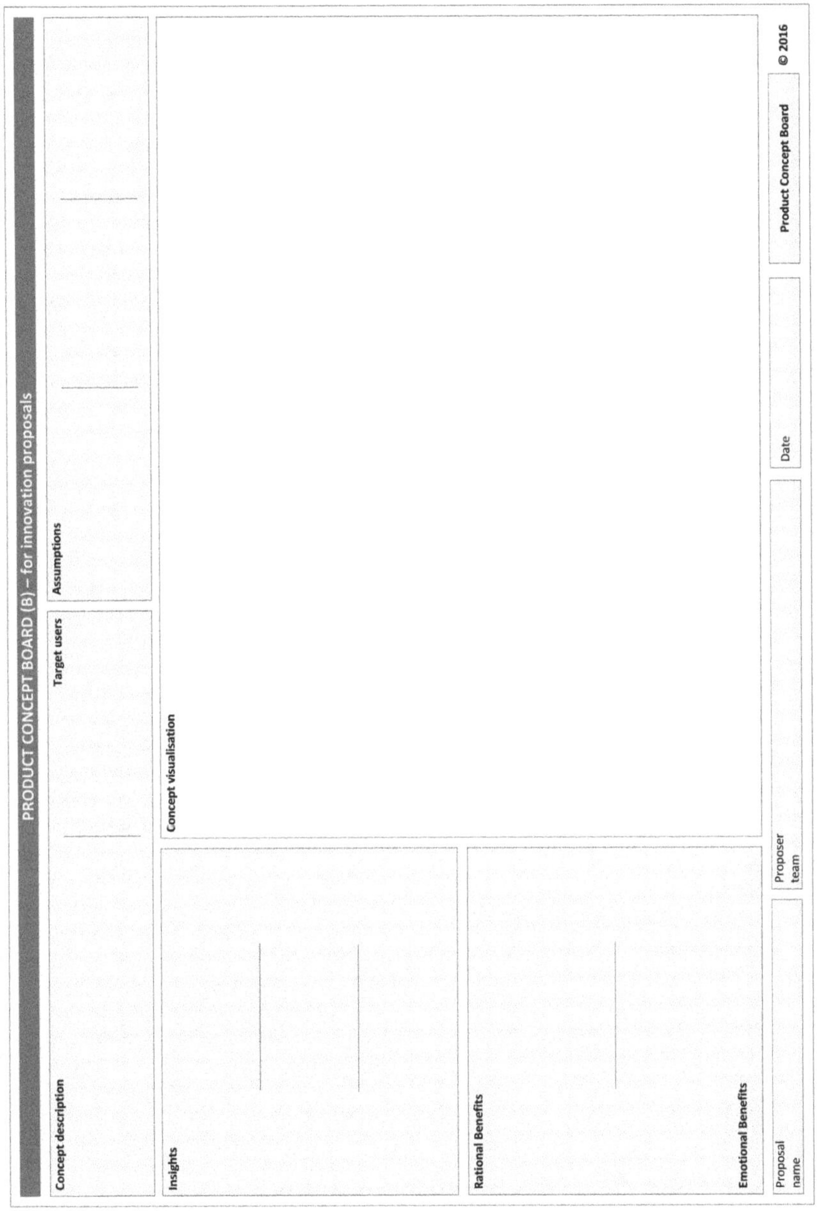

FIGURE 7.8 Product Concept Board (B) template.

PRODUCT CONCEPT BOARD (B) – for innovation proposals

Concept description

Target users

Assumptions

Concept visualisation

Insights

Rational Benefits

Emotional Benefits

Proposal name

Proposer team

Date

Product Concept Board

© 2016

FIGURE 7.9 Product Concept Board (B) example.

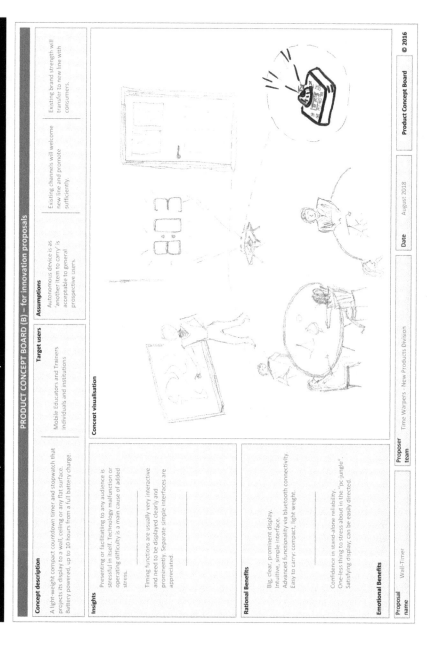

PRODUCT CONCEPT BOARD (B) – for innovation proposals

Concept description

A light-weight compact countdown timer and stopwatch that projects its display to a wall, ceiling or any flat surface. Battery powered, up to 10 hours from a full battery charge.

Target users

Mobile Educators and Trainers
Individuals and institutions

Assumptions

Autonomous device is as 'another item to carry' is acceptable to general prospective users.

Existing channels will welcome new line and promote sufficiently.

Existing brand strength will transfer to new line with consumers.

Concept visualisation

Insights

Presenting or facilitating to any audience is stressful in itself. Technology malfunction or operating difficulty is a main cause of added stress.

Timing functions are usually very interactive and need to be displayed clearly and prominently. Separate simple interfaces are appreciated

Rational Benefits

Big, clear, prominent display.
Intuitive, simple interface.
Advanced functionality via bluetooth connectivity.
Easy to carry: compact, light weight.

Emotional Benefits

Confidence in stand-alone reliability.
One-less thing to stress about in the "pc jungle".
Satisfying display; can be easily directed.

| Proposal name | Wall-Timer | Proposer team | Time Warpers - New Products Division | Date | August 2018 | Product Concept Board |

© 2016

Q How is it right to condense months of work into one poster, and hope to impress?

A Every innovative proposition must have a central, core value offering that is new. It also has many ancillary features; some are necessary, and others are superficial but nice. The core works without some of the others; nothing works without the core. Your first task is to make clear what is core. Describing the core does not mean there is nothing else. Everyone presumes you have volumes of design notes.

Secondly, too much information up front confuses any viewer. Introduce your proposition gradually and entice the viewer to explore further. This is the role of Concept Board (B).

Q We are working on a new HR on-boarding process. How could we visualise that?

A Everything can be represented visually, i.e. using symbols and effects other than text and numbers! For an on-boarding process, you might use a block diagram, a flow chart and/or a storyboard. It can be anything to help the viewer better understand and remember the concept at an overview level.

Q Given that we do not use all our information, how much content should we use, e.g. how many insights, assumptions and benefits?

A It depends on the project. For some projects, there is a clear, stand-out insight and a clear, stand-out benefit. If so, then just include them and nothing else. If they are strong enough on their own, do not risk diluting them with other less important material.

In other situations, you may need more of one or the other.

Always use as little argumentation as possible. One or two strong points easily trumps multiple weaker ones.

Shakespeare captured the above very succinctly in his play, *Hamlet*, when Queen Gertrude alludes to insincerity in a character with the words: 'The Lady doth protest too much, methinks'.

Try not to protest too much, and you will avoid appearing to have only weak arguments, with the added value of maintaining maximum clarity.

The Business Case Dashboard provides an overview of the business case surrounding the business proposal. It complements and supports the Product Concept Board (B).

As the project crystallises to a solid business proposal requiring further investment and strategic support, your priority must be to persuade senior executives and others around you of its potential value to the company. In addition, often you may need to reconfirm in your own mind the essential justifications for progressing further.

Throughout *Validate* you have developed and partially confirmed your thoughts (assertions) surrounding the *viability, feasibility* and *strategic suitability* of your proposal. The Business Case Dashboard packages these in dashboard format for easy overview. It plays an important role in persuading investors and procuring funding for the *limited launch* phase, described in this chapter, the purpose of which is to continue refining through testing in real market settings.

As you proceed through the experiments and adjustments of the *limited launch*, continue to review its contents and prepare for full launch with up-to-date revisions of the Business Case Dashboard in which you have an increasingly high level of confidence based on your accumulating evidence.

Step by step

1. **Allocate responsibilities and prepare data**
   Allocate responsibilities for preparing content for each of the five sections of the Dashboard template on the next page. Preferably, the preparation work is done by team members. If a section requires specialist input, for example finance or marketing, allocate a team member to liaise and drive this.

   Allow one week, no more. You must work off your team's best knowledge of the moment. You are not commissioning a separate new research project. It won't be perfect at this (early *Execute*) stage. The purpose of *Execute* is to make it better!

2. **Present the content of each section**
   Convene a whole-team meeting, to last 3–4 hours. Arrange 7–10-minute presentations on each section of the Dashboard, given by the responsible team member, where the presentations focus on:

   • the final content for the section, and
   • key doubts or uncertainties.

   Keep focus. Discourage deviating discussion.

   Allow a ten-minute discussion, mostly for clarification, after each presentation. Do not make 'corrections' to the content of the presentations yet.

3. Review each section

Taking each section in turn, discuss it freely and amend the content as necessary. Pay particular attention to the questions:

Does it represent our best understanding?
What are the significant uncertainties?
Is it clear to an uninitiated viewer?

4. Complete and revise regularly

Transfer the finished content to a good-sized template as a poster (e.g. A1 size, 2 ft × 3 ft or similar). Do not overfill the template; keep it easy to absorb.

Be sure to revise this regularly as you encounter new learning throughout the *limited launch*.

BUSINESS CASE DASHBOARD template

FIGURE 7.10   Business Case Dashboard template.

BUSINESS CASE DASHBOARD – for innovation proposals

Use with Product Concept Board (B)

Is it financially sustainable and scalable?

	Year 1	Year 2	Year 3
Revenue (€k) (or cost saving)			
Cost of Sales (€k)			
Gross Margin (%)			
No. of units (Volume)			
Average unit price (€)			

What investment is required (€)?

TOTAL (€k)

in the period   from: to:

costing off

Product development ...

Manufacturing/product realisation ...

Market launch ...

PAYBACK PERIOD

IRR

How does the project align strategically with our organisation?

Why is it suitable for our organisation?
(given our mission and strategy)

What significant organisational changes are required?

Why should we undertake it now?

What new strategic partnerships are necessary?

What technology will be used and what is the development risk?
Platform and system technology

What is the key technology(-ies) to be used?

Why is it the most suitable?

Is it readily available to the firm?

Is the proposed technology? ...

	routine for the firm	new to the firm	stretch for the firm	new to the world

How will it be branded and positioned in the market?

Positioning Statement

"For (target customers)

who are dissatisfied with (statement of the need or opportunity)

our product (name)

is a (product category)

that provides (key benefit)

Unlike (product alternative)

our product (statement of primary differentiation)

What is the foundational insight (into the customer's mind) that drives this positioning?

How will it relate to (or impact on) the firm's existing brand positioning?

Development

What is the timeframe to deployment?   .... months

Where will it be done?   in-house   outside
partner/contractor

What is the risk?   (very low)   (very high)
Time-to-deployment   1 2 3 4 5
Performance   1 2 3 4 5
Cost   1 2 3 4 5

Date

Business Case Dashboard

© 2016

Proposer team

Proposal name

BUSINESS CASE DASHBOARD example 3/4

FIGURE 7.11 Business Case Dashboard example.

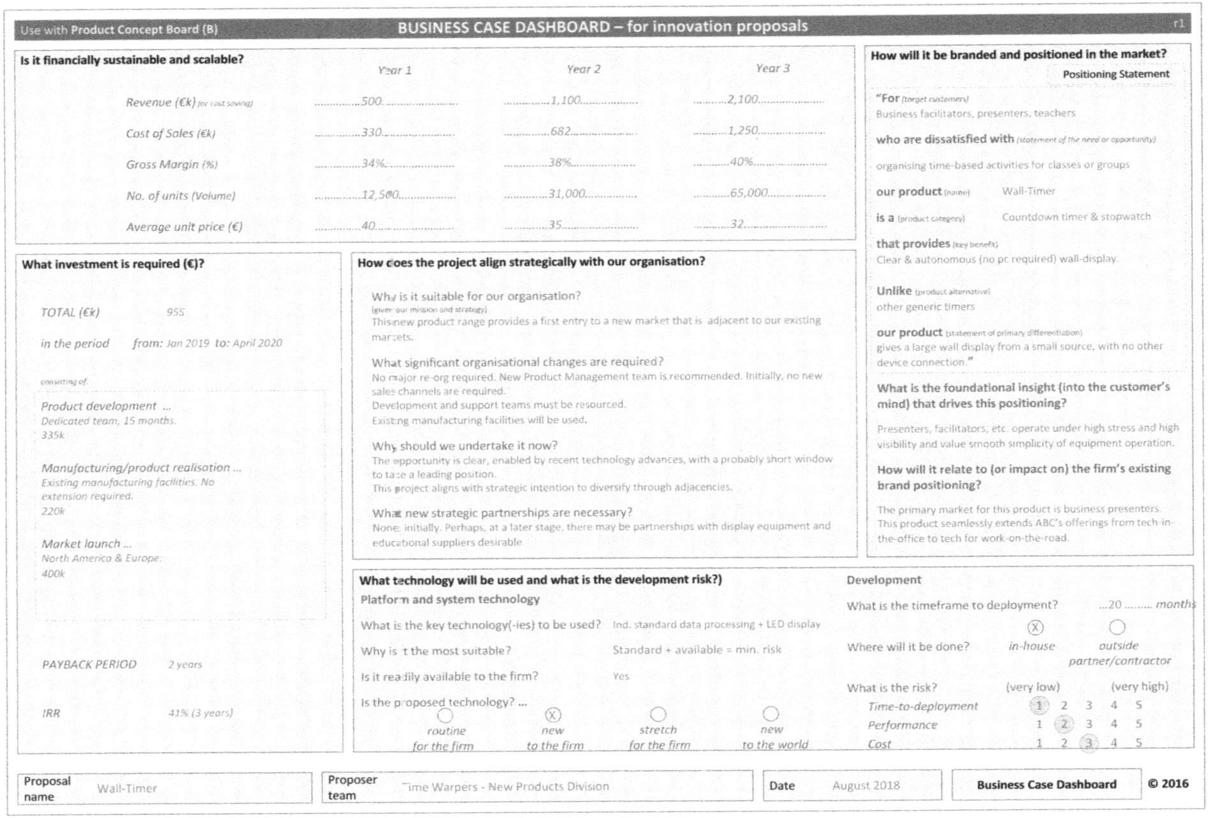

Use with Product Concept Board (B)

BUSINESS CASE DASHBOARD – for innovation proposals

Is it financially sustainable and scalable?

	Year 1	Year 2	Year 3
Revenue (€k) *(w/ cost saving)*	500	1,100	2,100
Cost of Sales (€k)	330	682	1,250
Gross Margin (%)	34%	38%	40%
No. of units (Volume)	12,560	31,000	65,000
Average unit price (€)	40	35	32

What investment is required (€)?

TOTAL (€k) 955

In the period from: Jan 2019 to: April 2020

comprising of:

Product development ...
Dedicated team. 15 months.
335k

Manufacturing/product realisation ...
Existing manufacturing facilities. No
extension required.
220k

Market launch ...
North America & Europe.
400k

PAYBACK PERIOD 2 years

IRR 41% (3 years)

How does the project align strategically with our organisation?

Why is it suitable for our organisation?
(given our mission and strategy)
This new product range provides a first entry to a new market that is adjacent to our existing markets.

What significant organisational changes are required?
No major re-org required. New Product Management team is recommended. Initially, no new sales channels are required.
Development and support teams must be resourced.
Existing manufacturing facilities will be used.

Why should we undertake it now?
The opportunity is clear, enabled by recent technology advances, with a probably short window to take a leading position.
This project aligns with strategic intention to diversify through adjacencies.

What new strategic partnerships are necessary?
None, initially. Perhaps, at a later stage, there may be partnerships with display equipment and educational suppliers desirable.

What technology will be used and what is the development risk?

Platform and system technology

What is the key technology(-ies) to be used? Ind. standard data processing + LED display

Why is it the most suitable?	Standard = available = min. risk		
Is it really available to the firm?		Yes	

Is the proposed technology? ...

routine	new	stretch	new
for the firm	to the firm	for the firm	to the world
	⊗	○	○

Proposal name Wall-Timer

Proposer team Time Warpers – New Products Division

Date August 2018

How will it be branded and positioned in the market?

Positioning Statement

"**For** *(target customers)*
Business facilitators, presenters, teachers

who are dissatisfied with *(statement of the need or opportunity)*
organising time-based activities for classes or groups

our product *(name)* Wall-Timer

is a *(product category)* Countdown timer & stopwatch

that provides *(key benefit)*
Clear & autonomous (no pc required) wall-display

Unlike *(product alternative)*
other generic timers

our product *(statement of primary differentiation)*
gives a large wall display from a small source, with no other device connection."

What is the foundational insight (into the customer's mind) that drives this positioning?
Presenters, facilitators, etc. operate under high stress and high visibility and value smooth simplicity of equipment operation.

How will it relate to (or impact on) the firm's existing brand positioning?
The primary market for this product is business presenters. This product seamlessly extends ABC's offerings from tech in-the-office to tech for work-on-the-road.

Development

What is the timeframe to deployment?		20	months

Where will it be done?	in-house	⊗	outside partner/contractor	○

What is the risk?

	(very low)				(very high)
Time-to-deployment	①	2	3	4	5
Performance	1	②	3	4	5
Cost	1	2	③	4	5

Business Case Dashboard

© 2016

Q Are there other topics that might belong in a Business Case Dashboard?

A Yes, there probably are. For example, we have not dealt with competitive response. How will incumbent competitors respond to our entry with a new offering? There are certainly many more.

We encourage you to add or subtract to the Dashboard according to the norms and expectations in your own firm.

But, resist creating a forest of data. Find the balance between comprehensiveness and accessibility. The purpose of this Business Case Dashboard is to give a clear overview to people who are busy.

Q How often should we update the Dashboard?

A As soon as you get new information about your project's prospects, consider the consequences and update it as appropriate. Do this with positive and negative information, and let the Dashboard be a pointer for your best estimate at all times.

Q What is the point in making estimates of financial data, for instance, when we really are not sure?

A How many units do you think you will sell next year? Our own guess may be 10,000, or 100,000. But that is a useless contribution, because we do not know anything about the product in question or the market you plan to serve.

Nobody else knows as much as you do about your proposed product and market! Your guess is based on immersion and experience. Others' guesses are pure speculation or based on second-hand guessing from information supplied by you. Your estimate may be off, but it will have some basis from your project researches, which can be used as a reference. Its quality will improve throughout your Execute testing.

The point is to give your best estimate, which is based on your project immersion, and supplement it with your estimate of uncertainty. Your prospective investor must have some basis for decision, and you are the expert to supply it.

EXECUTE TEST PLAN method 1/4

Rigorous marketplace testing of the limited launch requires careful and astute planning.

You must carefully plan how you are going to test each assumption. As we have repeatedly said, it involves a considerable amount of ingenuity to get the most learning bang from a limited number of bucks.

For each test, define the measure that will tell you whether the test has been passed. If you don't objectify the metrics at this stage, it is probable your emotional attachment to the project will cause you to be too forgiving of failures or 'almost' successes when they inevitably occur.

With clear and dispassionately considered metrics of success, resolve not to proceed with the project until each metric has been surpassed – or you can rationally justify an alternative metric or plan. For example, for an assumed price point of €1,500, your metric might be that you can get at least 20 paying customers in your first three months, and 40 paying customers in your second three months. If you fail to achieve your predicted numbers by a substantial gap, then you must seriously question your understanding of the value proposition you offer, its value to the target customers you have chosen, the efficacy of the delivery model you have used, or all three. Something is wrong, and it may be fatal! You have to triage it, identify the problem and test again, urgently.

Step by step

1. **Gather priority assumptions and the team**
   Create a listing of the most important remaining assumptions, most likely from a recent Assumption Mapping exercise. Then, convene a two- to three-hour meeting of the team. Display the Priority Assumption List for all to see.

2. **Devise experiments**
   Divide the participants into pairs and allocate each assumption on the list to one pair of participants, until all are allocated. Participant pairs then consider each assumption in turn, allocating approximately ten minutes to each. Complete the sections in the template overleaf.

   - **Related hypothesis to test:** Drill down into each assumption to see what is the underlying testable hypothesis(es). There may be more than one. A validly stated hypothesis is one that is capable of being proved wrong in an experiment.
   - **Experiment:** Devise a suitable experiment to test the hypothesis. Apply imagination and ingenuity to devise a suitable experiment that delivers maximum learning with minimum cost and time.
   - **Metrics:** Clearly specify how you will measure success and failure in the experiment.

3. **Discuss and agree with the team**
   Bring the team together again, and go through each experiment in turn. The experiment creator (pair) explains the hypothesis, proposed experiment and metrics. The team discusses the proposal and seeks to adjust it if necessary before agreement.

4. **Allocate responsibility**
   Allocate the experiment to an 'owner' within the team. The owner is tasked with managing all aspects of the experiment and reporting back findings within a certain timeframe. A useful template for planning and recording each experiment is shown on the next page: Learning Experiment Canvas.

FIGURE 7.12 (a) Execute Test Plan and (b) Learning Experiment Canvas templates.

EXECUTE Test Plan

Assumption	Related hypothesis to test	Experiment	Success measure
Desirability			
Viability			
Feasibility			
Strategic Suitability			

Learning Experiment Canvas. Plan your learning experiment and prototype to explore/validate your concept assumption.

Project name	Concept summary
Problem statement	
What is the main problem statement that gives rise to the concept?	

What is the **assumption** being tested? (an unambiguous declarative statement in <20 words)	At what stage is the concept? (place X along the line)
e.g. The user will...	Early stage / viable answer ——————— later stage / final decision

Describe the experiment	Outline the prototype you will use
e.g. Which users will be targeted? Who will participate from innovation team? What context/setting? What props/actors? What activity/interaction will be created?	e.g. What is the physical form? (storyboard, sketch, video, cardboard model, role play, skit, other) Sketch it.

Expected results from the experiment?	Realised results? [later]	Learning & Outstanding issues? [later]
e.g. User behaviour? User attitude? Did conversion get to convince? Later stage concepts involve?	e.g. User behaviour? User attitude? Did test design reject assumption, etc? Yes, stage concepts involve?	

Next steps [later]	Date concluded	© DEVITT Design Innovation
What are the next steps to pursue? Does it change? Less, more, more?		

FIGURE 7.13   Execute Test Plan example (extract).

EXECUTE Test Plan

Assumption	Related hypothesis to test	Experiment	Success measure
Desirability			
D1. The saddle's characteristics provide great benefits for the 'eventing' sector by its modular design and suspension characteristics.	The modular design of the cantilevered saddle will appeal to semi-professional eventing riders, as they need different types of saddles for different parts of the event. The modularity offers a more cost effective solution to multiple saddles.	Demonstrate the saddle to a variety of horse disciplines. Assess the reaction of eventers compared to other disciplines.	Eventers must show a tangible excitement concerning the saddle's prospects and strong intentions, with some commitments, to purchase the saddle.
...			
Viability			
V1. The best sectoral launch target is 'eventing', which has greatest pick-up rate prospects and price per performance potential.	(a) Our target retail price is reasonable for this market. (b) There is a fast rate of adoption of novelties in this market.	Arrange for >10 professional and hobby event riders to test the saddle over a period of multiple weeks	**(a) Greater than 70%** of riders who have trialled the saddle confirm their willing-to-pay price exceeds our target retail price. **(b) ...**
...			
Feasibility			
F1. Subcontract manufacturer and assembly suppliers can ramp up volume quickly.	(a) If our demand increases, the manufacturer and assembly suppliers will be able to respond to extra demand within 1 month. (b) Alternative/additional suppliers are available.	Conduct potential suppliers' audit. Space, machine, labour capacity. Critical materials: global market availability. Supplier history of flexibility with other customers.	No major flaw found. At least two alternatives for each of our existing key suppliers is currently available.
...			
Strategic Suitability			
S1. Our key investors are still behind us	When presented with successful 'Execute' phase outcomes, investors will be willing to provide the needed support.	Prepare an investment plan based on successful Execute outcomes. Ask them to commit!	At least 50% of investors (if n>6) or oversubscription to 150% promised (whichever is higher).
...			

Q What is the connection between the Test Plan and the Learning Experiment Canvas?

A The Learning Experiment Canvas is for detailed description and tracking of an individual experiment. The Execute Test Plan is an overview of all the priority experiments that are foreseen at this time.

Q Is it necessary to have a separate test for each assumption?

A No. Often, it is possible to combine testing of more than one assumption in a single experiment. If this can be done efficiently and effectively, that is great.
Be careful, however. The promise of efficiency by combining experiments might be superseded by the extra complexity of the experiment and possibly, even worse, corruption of the results and their interpretation by cross-contamination effects.

Q What is the difference between 'test' and 'experiment'?

A Apologies for any confusion. The difference is small and we don't mean to make a big deal of it. Mostly, we use the terms interchangeably. However, perhaps we are subconsciously trying to indicate a hybrid, as follows.
'Test', to our ears, implies a quicker, more direct, straightforward evaluation. 'Experiment' implies more formality, preparation, set-up, formal measurement and analysis.
In fact, what we are mostly seeking here is a judicious combination of the two, i.e. a balance that achieves speed, with clear definite results as much as possible.

BIBLIOGRAPHY

Blank, Steve (2013). *The Four Steps to the Epiphany* (5th edn), K & S Ranch.

Liedtka, Jeanne (2011). Learning to Use Design Thinking Tools for Successful Innovation. *Strategy & Leadership*, Vol. 39, No. 5, pp. 13–19.

Liedtka, Jeanne and Ogilvie, Tim (2011). *Designing for Growth*. Columbia Business School.

Maurya, Ash (2010). *Running Lean*. O'Reilly Media.

Maurya, Ash (2016). *Scaling Lean*. Portfolio/Penguin.

Osterwalder, Alexander and Pigneur, Yves (2010). *Business Model Generation*. John Wiley and Sons.

Ries, Eric (2011). *The Lean Startup*. Portfolio Penguin.

CHAPTER 8

Epilogue

We conclude this book with a short summary of its highlights, which we view as the key contributions that design innovation makes to innovation in business and other organisations.

In its 2017 Innovation Benchmark report published in Spring 2018, business consulting company PWC reported that 59 per cent of its globally surveyed companies use design thinking as an innovation methodology (PWC, 2017). It is not surprising that many of these companies wish to benefit from the enhancement in innovation and overall business performance that design thinking promises. It is also not surprising, especially in view of the recency of its introduction to mainstream business, that many of the implementations of design thinking are incomplete, superficial or low level.

In October 2018, another consulting company, McKinsey, published a report on the Business Value of Design (Sheppard et al., 2018). In their research they found, when firms implement design thinking well and comprehensively across the corporation, significant business performance improvements are evident. Their report tracks data over five years and shows these high-performing design thinkers have average annual Total Returns to Shareholders of 21 per cent and Revenue Growth of 10%. These numbers are large increases on industry benchmarks of 12–16 per cent and 3–6 per cent, respectively.

The message is clear. Design thinking helps innovation. And, design thinking that is implemented well can be a game changer for businesses.

In this book, we consider design thinking as an integral part of an innovation process that we call Design Innovation, and we focus on corporate innovation, although the principles and benefits apply to many other spheres.

FRAMEWORK AND PROCESS

We offer a framework model of Design Innovation that also acts as a meta-process. The six stages of the ARRIVE model describe the full life cycle of innovation, from early audit and research through to execution and deployment.

FIGURE 8.1 The ARRIVE framework for Design Innovation.

audit
What is known?

research
Find out more

reframe
Change Perspective

ideate
New Ideas & Concepts

validate
Test with users

execute
Develop, deploy & scale

Why a process? We know that the notion of a process as a tightly controlled way of doing business does not sit easily with the message of free creativity encouraged by many firms' innovation teams, especially at the earlier stages of innovation. However, the process view is valuable for three reasons (Devitt et al., 2017):

1. Business organisations are multifunctional and have requirements to coordinate across functions and to perform consistently. Their natural ways of doing this are process based. Hence, organisations that wish to bring on board the benefits of design thinking in order to stimulate and sustain innovation need a design thinking process, or at least an overarching or meta process, that serves to provide a choreography for the various methods and behaviours that make up design thinking. Processes are part of the natural landscape of businesses.

2. In too many cases, design thinking is practised as a varied collection of disparate methods. The fragmentation of design thinking into myriad methods allows innovation teams to lose their way and, being lost, rush prematurely towards ground that is more familiar. An example of this is a common tendency to skip too quickly to solution development before properly researching the problem space or before considering discoveries gained from research by synthesising insights and reframing. Engagement with disparate individual methods is transactional and short-lived in nature instead of integrating them as a standard coherent practice. The ARRIVE meta-process provides a unitary narrative of the innovation journey from beginning to end, so that practitioners can know where they are at any stage.

3. A process impacts further than one or more project, task or activity. Process encourages repetitive behaviour, and good repetitive behaviours change mindsets and cultures for the better (Lehman et al., 2004).

Finally, while we use the word process, we do not mean to say that a slavish stage-by-stage adherence must apply in all cases. It need not be all or nothing, and we have described many circumstances where 'all' is not possible. If you choose to skip parts of the process, be conscious that there is an increased risk to the project. We regard the process as a de-risking tool for transformational innovation; as you use more of it, a successful outcome to your project is more likely. As you use it less, the likelihood of mistake and unsuccessful outcome increases, though success is not impossible.

HUMAN-CENTRED INNOVATION

Here is a contentious statement for some people: 'Machines don't have problems!'

We are happy to assert this because machines are inanimate objects (for the moment, at least) with no inherent desires or aspirations. Humans, on the other hand, are full of aspirations and desires for particular outcomes. When humans have problems, they arise because they experience barriers to achieving their desired outcomes.

Machines, circumstances, systems, strategies and government policies have features. A problem crystallises only when a human tries to interact with them to achieve a desired experience. Hence, you cannot understand a problem properly without considering the human element. Too many innovating firms attempt to include the human element minimally or superficially. Design Innovation is distinctive because it starts and ends with the human considerations.

In the early stages of the innovation, we research the human element of the problem space in detail (Research, chapter 3). We use various ethnographic methods and tools to get an understanding of users and other stakeholders, which goes far deeper than the superficial. This informs our understanding of the problem and the approach to resolving it that we adopt with Reframe (chapter 4).

Then, we ideate for a range of possible solutions and immediately bring our raw ideas and early concepts back to the user for feedback and advice on how to proceed. The co-creating interaction with users – testing, feedback and iterating – continues all the way through concept development and business model development until full-scale launch.

Human-centred innovation means taking guidance from users on the nature of the problem, and continuing advice from users on your progress in solving it. Staying with the user means you stay with the problem at hand and are less likely to stray into problem territories of your own imagination.

DVFS

Since Tim Brown (2009) introduced the triple attributes of a successful project they have become widely recognised. Brown said a successful innovation project must be *desirable, viable* and *feasible*. The succinct truth and usefulness of this concept is reflected by its popularity, and we applaud it.

Over many years, however, we have often found something missing. We have regularly encountered projects that are indeed desirable, viable and feasible by an objective evaluation, yet they do not come to fruition, and they do not repay early investment. Usually, this is due to a deficiency in what we term *strategic suitability*. For a project to thrive within an organisation or even with a sole entrepreneur, it must align with the strategic resources, aspirations and capabilities of the promoter firm or individual. It must be high enough on the firm's priority. It must match the firm's risk appetite and strategic direction. It must have a senior executive champion. The required investment funding must be neither too high nor too low. In chapter 1, we described the DVFS quartet of criteria for a successful project in a firm, which relate to the four principal dimensions of the problem space, viz. *user, market, technology* and *firm*.

FIGURE 8.2 Attributes of a successful innovation project *[Desirability–Viability–Feasibility–Strategic Suitability]*.

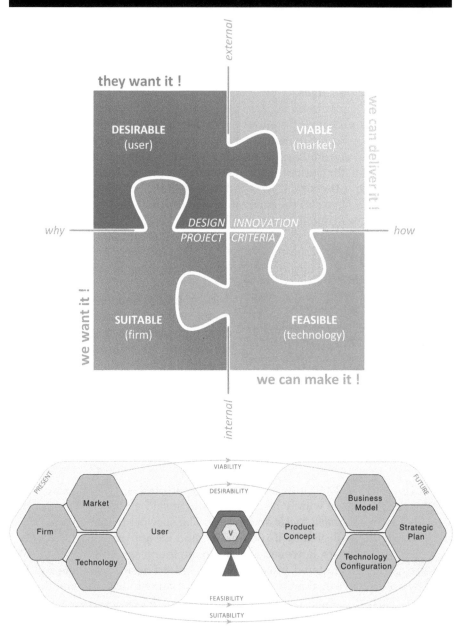

DESIGN THINKING DOES NOT STAND IN ISOLATION

We do not believe design thinking is uniquely distinctive, standing in isolation from all other approaches to improving business and innovation. Over the last 20 years, many new methodologies and movements have become popular, evolving from different sectors yet with common themes. Mostly, they attempt to redress the balance of the previous two decades away from the excessively rationalist and functionalist tendencies of technology-driven innovation employed by large global organisations with classical strategic frameworks. The flagship new principles are human understanding, experimentation and rapid, iterated learning.

Design thinking uses these principles to focus on exploring the problem space and ideating for new ideas in order to build new concepts for problem solutions. Other methodologies share the above principles, but with different contextual focus. For example, Lean Startup, described earlier, is focused on entrepreneurial new business development. Agile development is focused on developing products, especially software products, after the concept has been defined. Agile leadership describes a management and organisational style that facilitates rapid iterative learning throughout the organisation.

We believe design thinking is a particularly well developed methodology, ideal for getting to grips with the principles above, and it distinctively addresses the fuzzy front end of innovation This accounts for its current popularity, where organisations need to ensure they are doing the right project before they commit to developing it. It will be a measure of its success, probably not achievable before twenty years from now, when design thinking ceases to be a significant term in business literature. Through its own momentum and merging with other related movements as above, it will become the natural modus operandi of all high-performing organisations, as if there were no reasonable alternative. This is analogous to the evolution of the quality movement towards the end of the 20th century.

DESIGN THINKING = CRITICAL THINKING + GROWTH MINDSET

Design thinkers are 'smart creatives', a phrase coined by Google's recruiters. What makes design thinking such an effective and popular methodology is that it is a compelling integration of critical thinking and creative thinking. It practises and harmonises both of these critical skills as an overarching approach to a process fuelled by multiple specialist techniques and a growth mindset.

We have described extensively, in chapter 1, the parallels and symbiotic relationship between growth mindsets and design thinking practice. Both embrace the imperfection of our present state of knowledge and the value of validated learning, while continuously seeking out opportunities for such learning. Growth mindset is a characteristic of individuals that, through repeated practice, can spread to influence group and organisational culture. This is a facet of design thinking that appears to be well recognised by experts and non-experts alike.

Almost in contrast to the wide recognition of the above, too many critics and a good deal of practitioners of design thinking fail to see or practise its critical thinking elements, and hence many sceptical observers dismiss it easily as a vehicle of flaky creativity with limited practical or systemic relevance, as indeed it would be in circumstances where critical thinking were ignored.

Stephen Brookfield, long-term scholar of critical thinking and author of *Teaching for Critical Thinking* (Brookfield, 2012) provides a description of critical thinking as:

   i. uncovering and checking assumptions,
   ii. exploring alternative perspectives, and
   iii. taking informed action based on that enquiry.

The above mirrors exactly our ARRIVE design thinking meta-process, and that of others. New concepts are developed from creative efforts informed by user research. The creative leap incorporates assumptions, which are to varying extents supported by evidence, and reframing. These concepts are hypotheses that should not be fully executed until the underlying critical assumptions have been sufficiently assessed through prototyping and testing. Brookfield refers to assumptions such as these by the name 'causal assumptions'.

At the earlier research stage of an innovation project, design thinkers query the assumptions implicit in the status quo. For example, does a global accommodation company have to own its accommodation? Or, 15 years ago, did an airline have to provide free meals? Brookfield refers to assumptions such as these by the name 'paradigmatic assumptions'.

Besides the evaluation of assumptions, other elements of reasoning and justification that are additionally encompassed by critical thinking find many parallels in properly done design thinking practices. Critical thinking is essential for proper performance of a design innovation process, and two points deserve emphasis, as follows. If outsiders to the design thinking expert community recognise that critical thinking is integral, then many will respect design thinking more easily. In addition, good design thinking practice develops a heightened critical thinking facility in its practitioners, through repeated appropriate behaviours. This alone is a great systemic benefit for firms and individuals.

DESIGN THINKING OR DOING?

We conclude by considering a take-away point that is at the heart of design innovation. We use the term design innovation to refer to an innovation approach guided by design thinking.

Is it design thinking or design doing? Some people would prefer the latter because it places emphasis on action as in learning by doing, or experimenting. User research is also a field activity, not capable of being done satisfactorily while stuck in the office.

Repeated experimental actions, such as sketching, prototyping, experimenting and ethnographic observation, are all physical representations of the drive to learn, which is the essential characteristic of a growth mindset. The mindset drives the actions. Equally, repeated patterns of action build ways of thinking through habituation. Further, between thinking and action, there is clear evidence that the phenomenon of *visual thinking* is real. That is, the performance of 'visual' actions related to a thought process aids and alters the way we think and remember. Design thinkers build and act in order to think more clearly.

So, which comes first and is more important? It is like the chicken and egg. However, we like to stay with the term design thinking, because it is the *thinking* part that transcends the various disciplines. The form of experimental action may vary when designing a new software app compared to designing a new strategy or business model, but the thinking stays the same. Finally, when you try to spread the good word of design innovation across an organisation, action illustrates while thinking motivates.

BIBLIOGRAPHY

Brookfield, Stephen (2012). *Teaching for Critical Thinking*. Wiley.

Brown, Tim (2009). *Change by Design*. Harper Business.

Devitt, Frank; Ryan, Martin; Robbins, Peter; and Vaugh, Trevor (2017). Unlocking Design Thinking's Potential. *The ISPIM Innovation Summit – Building the Innovation Century*, Melbourne, Australia, 10–13 December.

Lehman, Darren R.; Chiu, Chi-yue; and Schaller, Mark (2004). Psychology and Culture. *Annual Review of Psychology*, Vol. 55, pp. 689–714.

Liedtka, Jeanne (2015). *Is Design Thinking the new TQM?* (report). https://www.iedp.com/articles/is-design-thinking-the-new-tqm/.

PWC (2017). *Reinventing Innovation: Five Findings to Guide Strategy through Execution. Key Insights from PwC's Innovation Benchmark*, PWC.

Sheppard, Benedict; Sarrazin, Hugo; Kouyoumjian, Garen; and Dore, Fabricio (2018). *The Business Value of Design*. https://www.mckinsey.com/business-functions/mckinsey-design/our-insights/the-business-value-of-design.

INDEX

Page numbers in **bold** and *italic* type refer to information in tables and figures respectively.

9 780367 618377